T0328296

Clinical Precision Medicine

Clinical Precision Medicine

A Primer

Edited by

Judy S. Crabtree, PhD

Associate Professor
Department of Genetics
Louisiana State University Health Sciences Center
New Orleans, Louisiana
United States

Scientific and Education Director
Precision Medicine Program
Louisiana State University Health Sciences Center
New Orleans, Louisiana
United States

Academic Press is an imprint of Elsevier
125 London Wall, London EC2Y 5AS, United Kingdom
525 B Street, Suite 1650, San Diego, CA 92101, United States
50 Hampshire Street, 5th Floor, Cambridge, MA 02139, United States
The Boulevard, Langford Lane, Kidlington, Oxford OX5 1GB, United Kingdom

Library of Congress Cataloging-in-Publication Data
A catalog record for this book is available from the Library of Congress

British Library Cataloguing-in-Publication Data
A catalogue record for this book is available from the British Library

ISBN: 978-0-12-819834-6

For information on all Academic Press publications visit our website at
https://www.elsevier.com/books-and-journals

Publisher: Andre G. Wolff
Acquisition Editor: Peter Linsley
Editorial Project Manager: Samantha Allard
Production Project Manager: Poulouse Joseph
Cover Designer: Alan Studholme

Typeset by TNQ Technologies

Contents

Contributors

Judy S. Crabtree, PhD
Associate Professor, Department of Genetics, Louisiana State University Health Science Center, New Orleans, LA, United States; Scientific and Education Director, Precision Medicine Program, Director, School of Medicine Genomics Core, Louisiana State University Health Sciences Center, New Orleans, LA, United States

Andrew D. Hollenbach, PhD
Co-director, Basic Science Curriculum, School of Medicine, Professor, Department of Genetics, Louisiana State University Health Sciences Center, New Orleans, LA, United States

Fokhrul Hossain, PhD
Postdoctoral Researcher, Louisiana Cancer Research Center (LCRC) and Department of Genetics, School of Medicine, Louisiana State University Health Sciences Center, New Orleans, LA, United States

Stephanie Kramer, MS, CGC
Certified Genetic Counselor, Center for Advanced Fetal Care, Clinical Assistant Professor, University of Kansas Health System, Women's Specialties Clinic, Kansas City, MO, United States

Samarpan Majumder, PhD
Instructor – Research, Louisiana Cancer Research Center (LCRC) and Department of Genetics, School of Medicine, Louisiana State University Health Sciences Center, New Orleans, LA, United States

Lucio Miele, MD, PhD
Director for Inter-Institutional Programs, Stanley S. Scott Cancer Center and Louisiana Cancer Research Center, Louisiana State University Health Sciences Center, New Orleans, LA, United States; Professor and Department Head, LSU School of Medicine, Department of Genetics, Louisiana State University Health Sciences Center, New Orleans, LA, United States

Fern Tsien, PhD
Associate Professor, Department of Genetics, Louisiana State University Health Sciences Center, New Orleans, LA, United States

CHAPTER

Cytogenetics in precision medicine

1

Fern Tsien, PhD

Associate Professor, Department of Genetics, Louisiana State University Health Sciences Center, New Orleans, LA, United States

Chapter outline

Cytogenetic techniques

Cytogenetic analysis traditionally involves G-banding (karyotyping). A higher resolution detection of constitutional and cancer-acquired chromosomal abnormalities can be achieved by combining the karyotype with molecular cytogenetic techniques such as fluorescence in situ hybridization (FISH) and microarray comparative genomic hybridization (aCGH). Each procedure has its advantages and limitations and can provide unique information regarding a patient's diagnosis and disease progression.

G-banding (karyotype) analysis

Karyotype analysis is highly efficient at identifying numerical chromosome abnormalities (e.g., trisomy, triploidy) and structural rearrangements (e.g., insertions, deletions, inversions, translocations) and is effective in uncovering cell population heterogeneity (Fig. 1.1). A limitation of this procedure is that aberrations ess than 1 Mb in size may be missed. Furthermore, to analyze metaphase chromosomes and identify rearrangements, living cells are required that are either actively undergoing cell division or induced to divide with the help of mitogens. Therefore, karyotype analyses cannot be performed on formalin-fixed paraffin-embedded (FFPE) tissue samples. Despite these limitations, G-banding is widely employed in both the research and clinical settings.

G-banding can be performed on almost any cell type that can be cultured (fresh live cells), including peripheral blood, solid tumors, bone marrow, skin fibroblasts,

Clinical Precision Medicine. https://doi.org/10.1016/B978-0-12-819834-6.00001-X

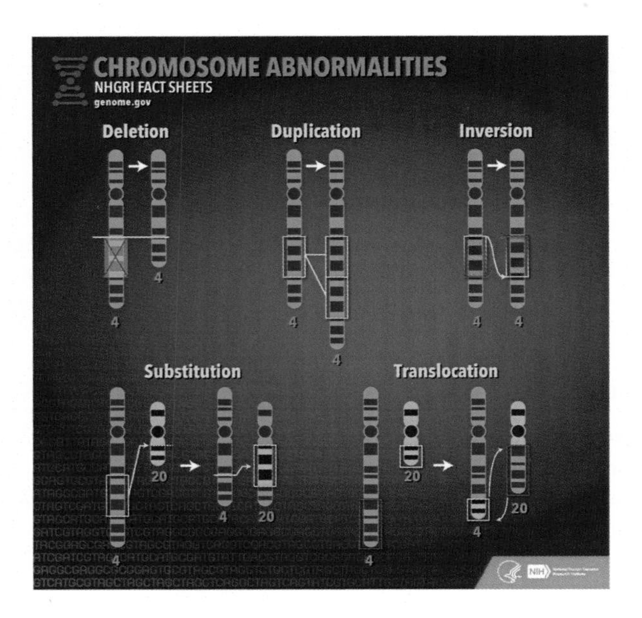

FIGURE 1.1

Structural chromosome abnormalities.

miscarriage material (products of conception), amniotic fluid, and chorionic villus sampling (CVS). Chromosomes are analyzed at the metaphase stage of mitosis, when they are most condensed and therefore more clearly visible. When a cell culture has reached an exponential phase with a high mitotic index, the cells are arrested at metaphase by disrupting the spindle fibers and preventing them from proceeding to the subsequent anaphase stage. The cells are treated with a hypotonic solution, preserved in their swollen state with a methanol-acetic acid fixative solution and then dropped onto glass microscope slides. The process of G-banding involves trypsin treatment followed by Giemsa staining to create characteristic light and dark bands.

Each individual chromosome can be identified by its distinct banding pattern and plotted on an ideogram or a map corresponding to the specific regions of each of the chromosomes. A classification system has been established in which each chromosome band is assigned a sequential number, starting from the centromere and increasing as one approaches the end of the telomere. All cytogenetic reports and publications utilize this International System for Human Cytogenetic Nomenclature (ISCN), which is continuously updated.

FISH is a procedure that combines basic principles of molecular biology and cytogenetics to evaluate chromosome abnormalities at a higher resolution than classic karyotyping. The procedure involves the hybridization, directly on the microscope slide, of a fluorescently labeled DNA probe to a complementary gene or chromosomal region. One of the main advantages of FISH is that it can be performed on mitotic and interphase cells, allowing for the analysis of archived tissue samples. Another benefit of FISH is that multiple probes of differing color can be implemented to concurrently analyze multiple genes, regions, or chromosomes, detecting translocations, amplifications, or other rearrangements diagnostic for a particular type of malignancy. FISH is ideally suited for the study of cancer-related chromosome instability (CIN), since it enables the analysis of cell morphology, and as a result, cell-to-cell heterogeneity. In general, both the number and size of FISH signals can be quantified, providing insight into the nature of a specific chromosomal aberration. One limitation of FISH is that the DNA probes relevant to a region of interest are not always commercially available. In addition to an ability to assess cell-to-cell heterogeneity, FISH can also evaluate CIN in samples isolated from the same patient at different time points to monitor disease progression and treatment response. Penner-Goeke et al. employed interphase FISH and assessed CIN in serial samples collected from women with ovarian cancer. They showed that an increase in CIN was observed in women with a treatment resistant form of the disease.

aCGH is a microarray procedure that can determine DNA sequence copy number changes throughout the entire genome. Fluorescently labeled DNA extracted from clinical samples is used as a probe. This DNA is mixed with normal labeled reference DNA and hybridized to a microarray chip. The laboratory utilizes specific computer software to view the ratio between the sample DNA (green) and the reference DNA (red), to determine gains or losses of DNA. Array CGH is used to detect amplifications, deletions, and chromosome gains and losses and is often implemented in cancer cytogenetic studies.

When aCGH is employed to compare the frequency of chromosomal imbalances in primary colorectal tumors and brain metastases, it can reveal a higher degree of sensitivity with regard to segmental aneuploidy in metastatic lesions. However, since aCGH employs pooled DNA samples isolated from large numbers of cells, it is incapable of measuring the level of cell-to-cell heterogeneity in chromosome number and structure that is characteristic of CIN. Another limitation of aCGH is that although it is efficient in detecting gains and losses of DNA material, it cannot detect rearrangements that involve inversions or translocations (i.e., the 9; 22 translocation in chronic myelogenous leukemia).

Chromosome abnormalities

Congenital chromosomal abnormalities usually result from abnormal nondisjunction or chromosome rearrangements during meiosis I or meiosis II (when the gametes are formed, prior to fertilization) and are sometimes parentally inherited. Alternatively, chromosome abnormalities may occur during the mitotic cell division of early development, resulting in mosaicism (two or more different cell populations). Chromosome abnormalities can be numerical (i.e., trisomy, monosomy, or polyploidy) or structural (i.e., deletions, duplications, inversions, insertions/substitutions, and translocations). In humans, chromosome abnormalities occur in approximately 1 per 160 live births, 60%–80% of all miscarriages, 10% of stillbirths, 13% of individuals with congenital heart disease, 3%–6% of infertility cases, and in many patients with developmental delay and/or other birth defects. Some disorders caused by congenital chromosome abnormalities can affect early prenatal development and may not be amenable to precision medicine.

However, **germline chromosomal abnormalities** or mutations can be associated with hereditary forms of cancer that may be responsive to precision medicine. For example, retinoblastoma patients who have the hereditary form (i.e., germline carriers of an *RB1* mutation or a chromosome 13q14 deletion) also have a risk of developing secondary malignant neoplasms such as osteosarcomas, soft tissue sarcomas, and melanoma. This risk of malignancy is maximized by external beam radiotherapy treatments, which is why these treatments are now avoided for patients with the hereditary form of retinoblastoma.

Acquired chromosomal abnormalities associated with tumor progression often result from nondisjunction or the formation of chromosomal rearrangements in postnatal tissues.

Malignant cells may include numerical and structural rearrangements similar to those observed in congenital cases. In addition, aberrations that are unique to malignant cells include **double minutes** and **homogeneously staining regions (HSRs)**. Double minutes are acentric (without a centromere) chromosomes composed of megabase-sized **extrachromosomal circular DNA (ecDNA)** elements frequently found in the advanced tumor stages of rhabdomyosarcomas, neuroblastomas, glioblastomas, and some hematological malignancies (Fig. 1.2). Multiple research groups have demonstrated that double minutes can readily evolve, therefore increasing tumor heterogeneity at a rapid pace. Turner et al. performed cytogenetic analyses and whole-genome sequencing of 17 different cancer types and analyzed the metaphase chromosome structure of 2572 dividing cells. They demonstrated that ecDNA was found in nearly half of the human cancers analyzed in their study. A variety of double minutes were observed that varied by tumor type and were rarely found in normal cells. Xu et al. suggest that various extrachromosomal driver oncogene amplifications (i.e., *MYC, EGFR, ERBB*) may enable tumors to adapt more effectively to different environmental conditions. This increases the probability that a subpopulation of cells will express one or more of these oncogenes, maximizing cell proliferation and survival and rendering these now malignant cells progressively more aggressive and difficult to treat over time (Fig. 1.3). Secondary somatic mutations such as insertions and deletions (INDELs), point mutations, and other rearrangements can occur on the double minutes after they are formed. One cell may contain multiple copies of double minutes with a specific genotype, and each one of these copies could acquire different mutations during one round of replication, leading to the accelerated evolution of double minutes. Therefore, the progression of double minutes appears to be independent of and faster than that of regular linear chromosomes. An understanding of the patterns of double

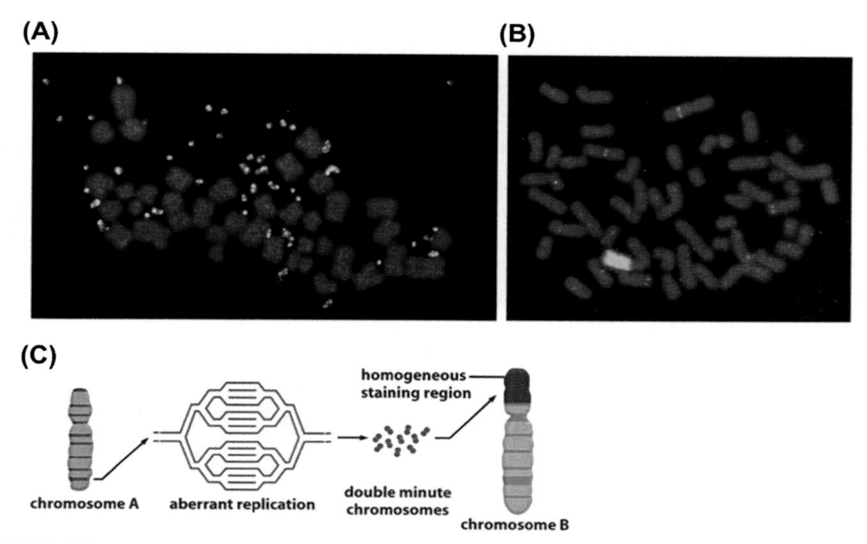

FIGURE 1.2

Fluorescence in situ hybridization of malignant cells using the *MYC* oncogene DNA probe demonstrating double minutes (A) and homogeneously staining regions (HSRs), (B). Double minutes are acentric chromosomes formed from aberrant DNA replication composed of megabase-sized extrachromosomal circular DNA (ecDNA). This ecDNA can then insert into linear chromosomes to form HSRs (C).

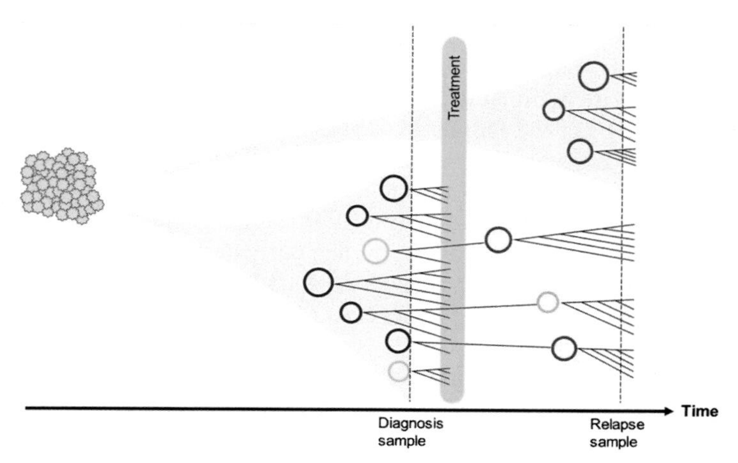

FIGURE 1.3

The cellular progression of double minutes.

minute evolution can lead to therapeutic solutions for tumors carrying these chromosomal rearrangements with specific oncogene amplifications.

ecDNA from double minutes can be spontaneously inserted back into linear chromosomes to form **HSRs.** These can be visually differentiated by routine G-banding or FISH; however, it is difficult to distinguish ecDNA from HSRs using next-generation sequencing.

The **Philadelphia chromosome** resulting from a 9;22 translocation is an abnormality that has made a significant impact in precision medicine. This translocation, resulting in a shortened derivative 22 chromosome, was the first genetic defect linked with a specific human cancer. It was discovered and described by David Hungerford from Fox Chase Cancer Center (previously the Institute for Cancer Research) in Philadelphia, PA, and Peter Nowell from the University of Pennsylvania School of Medicine; the derivative 22 chromosome was named after the city in which both facilities were located. In 90%–95% of patients with CML, the leukemic cells contain a translocation between chromosomes 9 (*ABL* tyrosine kinase gene) and 22 (*BCR*, breakpoint cluster gene). The resulting truncated chromosome 22 (Philadelphia chromosome) is designated Ph^1. At the molecular level, the Philadelphia chromosome contains the 5′ section of *BCR* fused with the 3′ end of *ABL* gene. Transcription and translation of the hybrid *BCR-ABL* gene produces an abnormal fusion protein that constitutively activates a number of cell functions that are normally inactive. Both ABL and the fusion protein BCR-ABL are tyrosine kinases: enzymes that attach phosphate groups to other proteins, often activating them. This unrestrained activation can increase the rate of mitosis and decrease the rate of apoptosis.

Genetic diagnosis of the 9;22 translocation is typically accomplished via karyotype analysis, FISH, and reverse transcriptase polymerase chain reaction (RT-PCR) of the patient's bone marrow specimen. All assays are recommended, since each provides specific disease information: karyotyping using G-banding not only reveals the presence or absence of the Ph^1 chromosome but also detects aneuploidy and rearrangements involving other chromosomes. FISH targeted to the *ABL* and *BCR* genes in interphase of cells can reveal the number of translocation events while simultaneously visualizing cell morphology (Fig. 1.4). Both techniques can provide information regarding clones (for example, the percentage of cells that contain a Philadelphia chromosome compared with cells with no translocation, and clonal evolution prior to and following treatment). RT-PCR can confirm the presence of *BCR-ABL* RNA transcript.

Philadelphia-positive leukemic cells have been successfully treated with the "first-generation" drug STI571 (Gleevec or imatinib mesylate). This molecule is a **tyrosine kinase inhibitor (TKI)** that fits into the active site of the ABL portion of the protein, preventing ATP binding. Without ATP as a phosphate donor, the fusion protein cannot phosphorylate its substrate. Although this does not eradicate CML cells, the risk of cells undergoing blast crisis does substantially decrease. A few patients can experience resistance or intolerance to imatinib due to rare point mutations/variants or alternative signaling pathways. "Second-generation" drugs

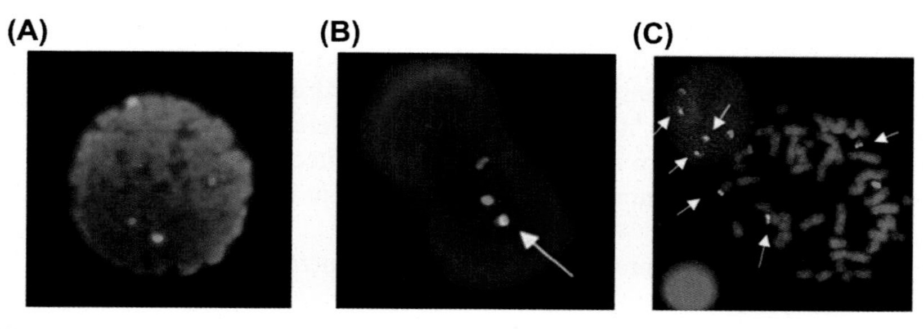

FIGURE 1.4

Detection of the ABL/BCR fusion by fluorescence in situ hybridization (FISH). FISH using the ABL probe on chromosome 9 (red) and the BCR probe on chromosome 22 in a normal cell (A), a cell positive for the 9; 22 translocation (B), and two cells (one in interphase and one in metaphase) from a patient with advanced CML with three ABL/BCR fusions in each cell (C).

that are currently FDA approved to overcome imatinib resistance include nilotinib, dasatinib, bosutinib, and ponatinib. Metaanalyses demonstrate that TKI discontinuation in chronic myeloid leukemia patients who have undergone prolonged complete molecular remission is feasible and safe. The duration of TKI therapy and of complete remission at the cytogenetic, molecular, and clinical level was inversely related to relapse rate after discontinuation. Relapse usually occurred during the first 6 months; however, almost all relapsing patients responded again to treatment after a TKI rechallenge. The 5%–10% of CML patients who do not exhibit the Philadelphia chromosome (Ph^1-negative) at diagnosis generally have a poorer response to treatment and shorter survival time compared with Ph^1-positive patients.

The 9;22 translocation can also be detected in acute lymphoblastic leukemia (ALL) and rare cases of chronic neutrophilic leukemia patients. At the protein level, different BCR-ABL transcripts can be generated, depending upon the precise location of the BCR gene breakpoint. Each breakpoint is associated with a different type of leukemia. The most common protein in CML is a 210-kD protein (P210); in ALL, a P190 is detected; in rare cases of chronic neutrophilic leukemia, a P230 isoform is created.

Chromosome instability (CIN) represents an overall increased rate of translocations, gains, losses, and other types of chromosome rearrangements, manifesting as cell-to-cell karyotypic heterogeneity and in many cases driving cancer initiation and evolution. Therefore, many current research efforts are aimed at identifying the etiological origins of CIN, developing novel therapeutics, and establishing its role in cancer pathogenesis and patient prognosis. CIN impacts overall genomic instability through the induction of simultaneous and ongoing copy number changes in a large number of genes (e.g., oncogenes, tumor suppressor genes, DNA repair genes, and apoptotic genes) that promote oncogenesis. The chromosomal gains, losses, and structural rearrangements associated with CIN in cancer progression promote the production of genetically distinct (i.e., heterogeneous) populations of daughter cells (clones). Thus, CIN during malignancy can increase intratumoral heterogeneity (ITH) that under certain conditions may confer a selective growth advantage (e.g., increased cell proliferation, metastatic potential, or drug resistance) to a select subpopulation of cells. Thus, it is not surprising that increased CIN is frequently associated with disease recurrence and poor patient outcome. Paradoxically, CIN can also be associated with more favorable outcomes but only in specific cancer types (e.g., CML).

In general, CIN can promote oncogenesis by increasing the rate at which key genes (e.g., oncogenes, tumor suppressor genes, DNA repair genes, and apoptotic genes) are gained, lost, or rearranged. Accordingly, CIN drives cellular transformation, cancer progression, ITH, multidrug resistance, tumor recurrence, and poor patient outcomes. Several research groups are developing a precision medicine strategy that selectively exploits the aberrant genes or pathways leading to CIN, specifically targeting a variety of cancers (at both primary and metastatic sites) to reduce and/or eliminate many of the off-target side effects associated with current chemotherapeutics. Many of these human CIN genes encode cell cycle checkpoints, sister chromatid cohesion, DNA replication and repair, centrosome duplication, mitotic spindle dynamics, kinetochore-microtubule attachment, and chromosome segregation. Currently, there are two fundamental strategies for exploiting CIN genes in cancer therapy: **CIN-reducing therapies** and **CIN-inducing therapies**. CIN-reducing slows down the rate of chromatin instability, typically inhibiting the chromosome missegregation or structural rearrangements seen in CIN-positive cancer cells. This strategy prevents the acquisition of additional chromosomal aberrations, minimizing intratumor heterogeneity and tumor cell adaptability and ultimately limiting cancer progression and drug resistance. Conversely, CIN-inducing therapies aim to exacerbate CIN, generating extreme levels of chromosome missegregation and/or DNA damage to induce apoptosis. While several in vitro studies have successfully employed chemical or genetic CIN-reducing and CIN-inducing approaches and have identified promising targets, many of these innovations have yet to be translated into the clinic.

Further reading

1. Bakker B, van den Bos H, Lansdorp PM, Foijer F. How to count chromosomes in a cell: an overview of current and novel technologies. *Bioessays* 2015;**37**:570–7. https://doi.org/10.1002/bies.201400218.
2. Chromosome Abnormalities. *National human genome Institute (NIH NHGRI)*. 2019. https://www.genome.gov/11508982/chromosome-abnormalities-fact-sheet/.
3. Driscoll D, Gross S. Prenatal screening for aneuploidy. *N Engl J Med* 2009;**360**: 2556–62. https://doi.org/10.1056/NEJMcp0900134.
4. Howe B, Umrigar A, Tsien F. Chromosome preparation from cultured cells. *J Vis Exp* 2014;**83**:e50203. https://doi.org/10.3791/50203.

5. Jusino S, Fernández-padín FM, Saavedra HI. Centrosome aberrations and chromosome instability contribute to tumorigenesis and intra-tumor heterogeneity. *J. Cancer Metastasis Treat.* 2018;**4**:43. https://doi.org/10.20517/2394-4722.2018.24.
6. Kaneshiro N, Zieve D, Reviewers. Down syndrome: trisomy 21. PubMed Health.
7. Kannan TP ZB. Cytogenetics: past, present and future. *Malays J Med Sci* 2009;**16**(2): 4–9.
8. Kong JH, Winton EF, Heffner LT, Chen Z, Langston AA, Hill B, Arellano M, El-Rassi F, Kim A, Jillella A, Kota VK, Bodó I, Khoury HJ. Does the frequency of molecular monitoring after tyrosine kinase inhibitor discontinuation affect outcomes of patients with chronic myeloid leukemia? *Cancer* 2017;**123**(13):2482–8. https://doi.org/10.1002/cncr.30608.
9. McGowan-Jordan J, Simons A, Schmid M, editors. *ISCN 2016, An international system for human cytogenomic nomenclature (2016) Basel.* Freiburg: Karger; 2016.
10. Lejeune J, G. M. Etude des chromosomes somatiques de neuf enfants mongoliens. *C R Acad Sci* 1959;**248**:1721–2.
11. Lepage CC, Morden CR, Palmer MCL, Nachtigal MW, McManus KJ. Detecting chromosome instability in cancer: approaches to resolve cell-to-cell heterogeneity. *Cancers* 2019;**11**(2):226. https://doi.org/10.3390/cancers11020226.
12. Ljunger E, et al. Chromosomal abnormalities in first-trimester miscarriages. *Acta Obstet Gynecol Scand* 2005;**84**(11):1103–7. https://doi.org/10.1111/j.0001-6349.2005.00882.x.
13. Mitelman F, Johansson B, Mertens F, editors. *Mitelman database of chromosome aberrations and gene fusions in cancer*; 2019. http://cgap.nci.nih.gov/Chromosomes/Mitelman.
14. Nowell P, Hungerford D. A minute chromosome in chronic granulocytic leukemia. *Science* 1960;**132**(3438):1488–501. https://doi.org/10.1126/science.132.3438.1488.
15. Nussbuam R, et al. *Genetics in medicine*. 7th ed. Philadelphia: Saunders/Elsevier; 2007. p. 76.
16. Penner-Goeke S, Lichtensztejn Z, Neufeld M, Ali JL, Altman AD, Nachtigal MW, McManus KJ. The temporal dynamics of chromosome instability in ovarian cancer cell lines and primary patient samples. *PLoS Genet* 2017;**13**:e1006707. https://doi.org/10.1371/journal.pgen.1006707.
17. Pierpoint M, et al. Genetic basis for congenital heart defects: current knowledge. *Circulation* 2007;**115**(23):3015. https://doi.org/10.1161/CIRCULATIONAHA.106.183056.
18. Reddy U, et al. Stillbirth classification—developing an international consensus for research: executive summary of a national institute of child health and human development workshop. *Obstet Gynecol* 2009;**114**(4):901–14. https://doi.org/10.1097/AOG.0b013e3181b8f6e4.
19. Rowley J. A new consistent chromosomal abnormality in chronic myelogenous leukaemia identified by quinacrine fluorescence and Giemsa staining. *Nature* 1973; **243**(5405):290–3. https://doi.org/10.1038/243290a0.
20. Thompson LL, Jeusset LM, Lepage CC, McManus KJ. Evolving therapeutic strategies to exploit chromosome instability in cancer. *Cancers* 2017;**9**(11):151. https://doi.org/10.3390/cancers9110151.
21. Turner KM, Deshpande V, Beyter D, et al. Extrachromosomal oncogene amplification drives tumour evolution and genetic heterogeneity. *Nature* 2017;**543**(7643):122–5. https://doi.org/10.1038/nature21356.

22. Xu K, Ding L, Chang TC, et al. Structure and evolution of double minutes in diagnosis and relapse brain tumors. *Acta Neuropathol* 2018;**137**(1):123–37. https://doi.org/10.1007/s00401-018-1912-1.
23. Zheng S, et al. A survey of intragenic breakpoints in glioblastoma identifies a distinct subset associated with poor survival. *Genes Dev* 2013;**27**:1462–72.

CHAPTER

Molecular genetics—the basics of gene expression

2

Andrew D. Hollenbach, PhD

Co-director, Basic Science Curriculum, School of Medicine, Professor, Department of Genetics, Louisiana State University Health Sciences Center, New Orleans, LA, United States

Chapter outline

A complex network of proteins, enzymes, and regulatory molecules orchestrates the expression of proteins required for the biological processes important for the proper functioning of humans. An equally complex network of regulation controls gene expression or the process by which the blueprint information stored in an individual's DNA is converted into these functional proteins. Because of this complexity, small variations or alterations at any step in the conversion of DNA into protein may have significant outcomes that contribute to the development of the diseased state. More importantly, although individuals may have what appears to be an identical pathological presentation of a disease, many times the underlying genetic alterations that led to the development of this disease may be different, which could greatly affect their responses to therapy. Therefore, it is important to understand the molecular genetics of gene expression to better understand how the treatments used in the clinic are based solidly in the molecular mechanisms of the disease.

In the late 1950s, Francis Crick proposed what would become known as the central dogma of molecular biology to describe the flow of gene expression from DNA through RNA and into protein (Fig. 2.1). In this central dogma, DNA serves as the template for the duplication of itself through a process called replication. The DNA also serves as the template for the production of the various forms of RNA (messenger RNA [mRNA], transfer RNA [tRNA], ribosomal RNA [rRNA], and

Clinical Precision Medicine. https://doi.org/10.1016/B978-0-12-819834-6.00002-1

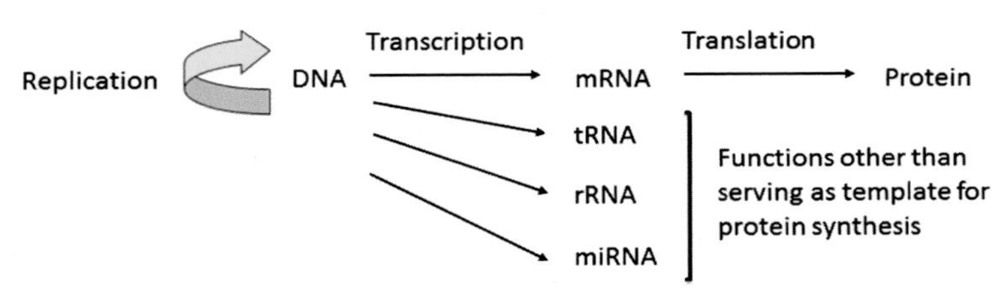

FIGURE 2.1

Schematic of the central dogma of gene expression.

microRNA [miRNA]) through a process called transcription. Finally, the mRNA serves as the template for the production of protein, which it does with the assistance of tRNA and rRNA, through the process called translation. The following discussion serves as a general overview of the process described by the central dogma, how regulation of this process occurs, and how mutations in the DNA can cause disruptions to biological processes, thereby leading to the development of the diseased state.

Replication

The ability of a cell to grow and divide requires that it be able to efficiently and faithfully reproduce its genome. It does this through a process known as replication. Replication uses both strands of the parent DNA as a template to generate two new copies of DNA through DNA synthesis, each of which will then be segregated to the daughter cells during mitosis. The process is initiated at origins of replication (ori) and proceeds in a bidirectional manner with each chromosome having multiple oris to achieve complete replication during S phase of the cell cycle. Furthermore, the cell has tightly controlled regulatory mechanisms to insure that an ori is used once and only once during replication.

For replication to proceed, the double-stranded DNA must first be unwound to generate two single strands that can each be used as the template for DNA synthesis. This localized unwinding generates a "bubble" that has two replication forks, each fork being described by the point where the helical double-stranded DNA and unwound single strands meet. Replication occurs on both forks with both strands of single-stranded DNA at each fork being used simultaneously. Synthesis is mediated through the enzyme DNA polymerase, which catalyzes the addition of nucleotides in the 5′ to 3′ direction. Because synthesis occurs in a specified direction, each strand uses a slightly different process to duplicate the DNA. One strand, known as the leading strand, is capable of synthesizing the DNA in a contiguous manner. In contrast, the second strand, known as the lagging strand, must synthesize the

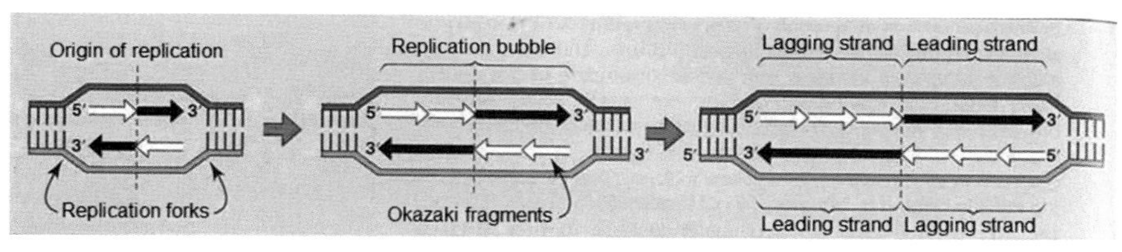

FIGURE 2.2

Illustration of a replication fork. Replication begins at the origin of replication and proceeds in a bidirectional manner, which results in the presence of two replication forks. The leading and lagging strands along with the presence of Okazaki fragments are indicated.

DNA in a noncontiguous manner through the generation of short pieces of DNA known as Okazaki fragments (Fig. 2.2).

Regardless of whether the synthesis is contiguous or noncontiguous, both use the same series of events and proteins to mediate the process. First, the enzyme DNA helicase unwinds the DNA at the replication fork to generate the two single-stranded DNA templates. Because DNA polymerase is unable to initiate synthesis in the absence of a primer, the enzyme RNA primase generates a short stretch of RNA that serves as the starting point for DNA synthesis. DNA synthesis continues until it reaches a subsequent RNA primer when the enzyme RNase H degrades the RNA primer, leaving a small stretch of single-stranded DNA. After DNA polymerase fills in this gap, the enzyme ligase covalently joins the final two nucleotides, thereby generating a continuous double strand of DNA.

Replication must be highly accurate to prevent the incorporation of incorrect bases. The incorporation of an incorrect base at this stage would result in a permanent change or mutation to the genome of the daughter cell. If this mutation results in the change of a key amino acid or alters a nucleotide in an important regulatory sequence, it could have serious detrimental effects that could potentially lead to a diseased state. Along these lines, DNA polymerase makes incorporation errors in roughly 1:100,000 bases. However, in addition to its polymerase activity, it also contains a proofreading exonuclease activity that is capable of removing incorrect bases as soon as they are detected, decreasing the error rate to 1:10,000,000 bases. Subsequent postreplication DNA repair enzymes further enhance accuracy to ensure the fidelity required for accurate replication of the entire genome.

DNA repair

Although replication has a relatively high level of accuracy, it is not 100% accurate, which allows for the incorporation of incorrect bases into the newly synthesized DNA strand. Furthermore, our DNA is under a constant onslaught of damaging agents, including UV radiation from ambient light or environmental agents that chemically alter DNA. The body has evolved several different mechanisms by which it is capable of repairing these misincorporations or environmentally altered DNA through the process of DNA repair.

1. Mismatch repair: If an incorrectly incorporated base is not corrected during the process of replication, it will result in a mismatch, or non-Watson-Crick base pairing, between the newly incorporated base and the template. In this instance, the Mut proteins recognize the mismatch and discriminate between the newly synthesized strand and the template strand. An endonuclease nicks the strand, an exonuclease removes the mismatched base, and DNA polymerase incorporates the correct base.
2. Nucleotide excision repair: UV radiation from ambient light may cause the covalent joining of two adjacent thymines (called "thymine dimers"). In this instance, the cells utilize nucleotide excision repair to remove the covalently joined bases, after which the gap is filled in by DNA polymerase.
3. Base excision repair: DNA bases such as cytosine may be deaminated, either over time or as a result of alkylating agents, to generate a uracil, a nucleotide present in RNA but not DNA. The incorrect base is recognized by a DNA glycosylase, which enzymatically removes the pyrimidine ring from the deoxyphosphate backbone, after which a specific endonuclease recognizes this altered base to excise and repair by DNA polymerase.

Transcription

The first step in the expression of a gene is the production of messenger RNA (mRNA) from the DNA template in the process known as transcription. Transcription generates the four different forms of RNA (mRNA, tRNA, rRNA, and miRNA), with mRNA serving as the template for subsequent translation into protein. All eukaryotic genes destined to become protein contain a core promoter, which is located immediately adjacent to the start of transcription. Because transcription can occur from either strand of DNA, the promoter is required to define the starting site and the direction of transcription (Fig. 2.3). The core promoter contains DNA sequences (e.g., the TATA box, BRE, DPE, etc.) that recognize and are bound by proteins known as general transcription factors. Eukaryotic RNA polymerases are unable to directly bind to DNA. Therefore, they must be recruited to the core promoter with the assistance of a host of general transcription factors, each of which is responsible for recognizing a different aspect of the core promoter and performing a specific function at the site of transcription initiation (Table 2.1). The binding of general transcription factors to core promoter elements creates the preinitiation complex that successfully recruits RNA polymerase to the transcription start site. This complex also defines what is known as basal transcription, or the minimal amount of transcription capable of occurring at a promoter in the absence of any additional regulation.

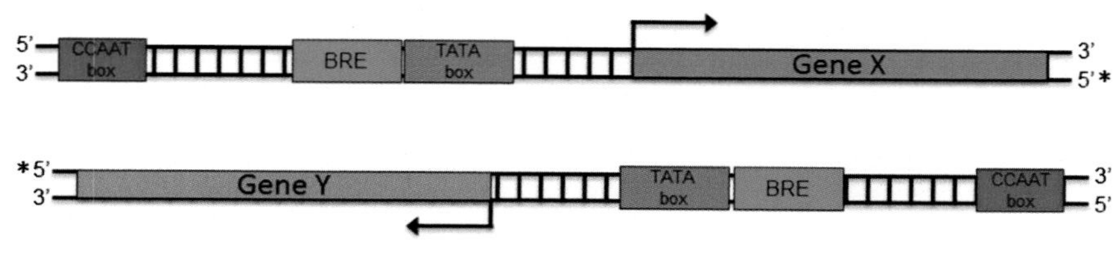

FIGURE 2.3

The role of the promoter in orchestrating the direction of transcription. The presence of regulatory sequences (CCAAT Box, BRE, and TATA box) facilitates the binding of transcription factors, which orchestrates the direction in which transcription will occur (indicated by the arrows). The placement of the regulatory sequences also directs which strand of DNA will be utilized as the template for transcription, which is indicated by the asterisks in both panels.

Once the preinitiation complex is formed at the transcriptional start site, the complex locally unwinds the DNA to create a single-stranded DNA template encompassing approximately 10—15 bases. The initial two ribonucleotides are brought to the active site where RNA polymerase catalyzes the covalent bond between them. After a period of abortive initiation, in which several short RNA molecules of less than 10 base pairs are synthesized and released, RNA polymerase breaks from the preinitiation complex to enter the elongation phase. During elongation, the polymerase continues to locally unwind DNA as it moves along the DNA strand utilizing nucleotides as dictated by the DNA template and Watson-Crick base pairing to polymerize the growing RNA strand. Elongation continues until sequences are encountered that signal the cleavage of the mRNA product from the growing strand, thereby promoting termination of transcription.

RNA processing

The mRNA generated from transcription is not the mature form that is ultimately utilized for translation into protein. A series of modifications must occur to this mRNA to generate its mature form. First, the mRNA must be capped on its free 5′-end. A series of enzymatic steps results in the covalent addition of a methylated guanine in a novel 5′ to 5′ linkage, a modification that will be used for the initiation of translation. In addition to 5′-capping, the mRNA undergoes polyadenylation on its 3′-end. A different series of enzymes recognize the unique polyadenylation signal present in the mRNA, cleaving the mRNA product from the growing strand followed by the covalent addition of approximately 200 adenines to the 3′-end.

The eukaryotic mRNA species that results from transcription, capping, and polyadenylation is not always a contiguous sequence with respect to what elements are required for subsequent translation. Instead the immature mRNA is often a series of coding sequences (exons) broken up by noncoding sequences (introns) (Fig. 2.4A).

Table 2.1 Listing of the general transcription factors, their subunit composition, and the general biological functions for the subunits, if a function is known.

Factor	Subunit Composition	Molecular Weight (kDa)	Function
TFIIA	TFIIAα	37	Required for activation
	TFIIAβ	19	Required for activation
TFIIB	——	35	Stabilized TATA-TBP interaction Recruits RNAPolIII-TFIIF
TFIIE	TFIIEα	56	Recruits TRIIH Stimulates TFIIH activities
	TFIIEβ	34	
TFIIF	RAP74	58	Facilitates RNAPolII promoter targeting Functional interaction with TFIIB
	RAP30	26	Destabilizes nonspecific DNA interactions
TFIIH	XPB, ERCC3	89	ATP-dependent helicase activity
	XPD, ERCC2	80	ATP-dependent helicase activity
	p62	62	
	p52	52	
	hSSI	44	
	Cyclin H	37	Kinase activity
	p34	34	Kinase activity
	Mat1	32	Kinase activity
	MO15	40	Kinase activity
TFIID	TBP	38	Binds to promoter TATA element
	TAF_{II}250 (TAF1)	250	Interacts with promoter (Inr, DPE) Protein kinase activity HAT activity
	TAF_{II}150 (TAF2)	150	Interacts with promoter (Inr, DPE) Required for initiator function
	TAF_{II}135 (TAF4)	135	Interacts with specific transcription factors
	TAF_{II}100 (TAF5)	100	Interacts with TFIIFβ
	TAF_{II}80 (TAF6)	80	
	TAF_{II}55 (TAF7)	55	
	TAF_{II}31 (TAF9)	31	Interacts with TFIIB Interacts with activation domains
	TAF_{II}30 (TAF10)	30	
	TAF_{II}28 (TAF11)	28	
	TAF_{II}20 (TAF12)	20	
	TAF_{II}18 (TAF13)	18	

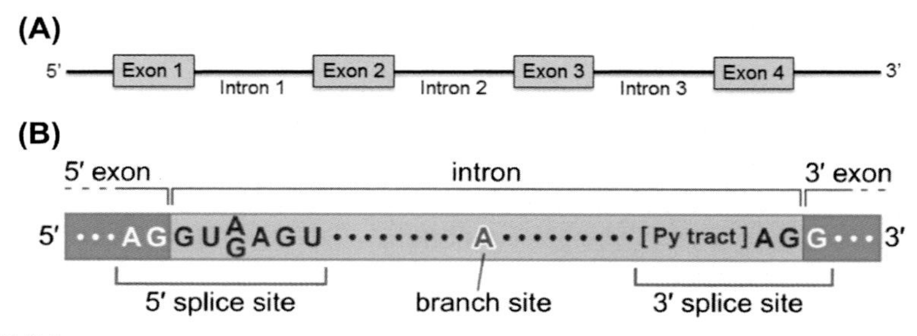

FIGURE 2.4

(A) Schematic of exon and introns structure along with (B) an illustration of conserved DNA sequences present at the 5′-splice site, 3′-splice site, and the branch site.

The number of exons and introns varies between genes with an individual gene having anywhere from one to several hundred introns with the size of the introns varying from several hundred to several hundred thousand bases. Regardless of the number or size of introns, this mRNA species, known as the pre-mRNA, must be further modified through the process of RNA splicing to remove these intronic sequences.

The spliceosome is the cellular machinery that catalyzes the chemical reaction required for splicing. It is a massive complex composed of nearly 150 proteins and 5 RNA species, known as small nuclear RNAs (snRNAs). The spliceosome comprises five individual complexes called small nuclear ribonuclear proteins, or snRNPs (pronounced "snurps") with each snRNP being composed of both protein and snRNAs. Through Watson-Crick base pairing between the snRNA present in the spliceosome and the mRNA to be spliced, the spliceosome recognizes explicit sequences located at the 5′-splice site (located at the junction between the 3′-end of the exon and the 5′-end of the intron), the 3′-splice site (located at the junction between the 3′-end of the intron and the 5′-end of the next exon), and a site within the intron, known as the branch site (Fig. 2.4B). Once recognized, the spliceosome brings these sites together and catalyzes a series of transesterification reactions to ultimately remove the intronic sequences and covalently join the exonic sequences. The process of covalently joining the exonic sequences must be individually repeated for each intron present in the gene. Once the mRNA has been capped, polyadenylated, and spliced to generate its fully mature form, it is then exported from the nucleus to the cytoplasm for translation.

Protein translation

The final step of gene expression is the translation of mRNA into protein. The process is called translation because the "language" of the genetic code (i.e., the nucleotide sequence) is translated into the "language" of proteins (i.e., amino acids). Translation requires four basic components:

1. The mature mRNA containing the sequence encoding the gene to be translated, processed, and exported to the cytoplasm, as described above. Only a portion of the mature mRNA is ultimately translated into protein. The stretch of mRNA that encompasses the initiating codon (i.e., the start codon) and the terminating codon (i.e., the stop codon) is known as the open reading frame. The region of the mRNA 5′ to the open reading frame is known as the 5′-untranslated region and serves to facilitate the initiation of translation. The region of the mRNA 3′ to the open reading frame is known as the 3′-untranslated region and contains recognition sequences for posttranscriptional regulation by miRNA (Fig. 2.5).
2. The tRNA, which has an amino acid covalently attached to it, recognizes the sequence of the mature mRNA, thereby bringing the correct amino acid to the ribosome for incorporation into the growing peptide chain. There are 61 individual tRNAs for 20 independent amino acids. Several amino acids can be attached to multiple tRNAs (e.g., leucine is attached to six independent tRNAs), thereby providing degeneracy in the genetic code.
3. The aminoacyl-tRNA synthetases covalently attach the correct amino acid to its corresponding tRNA. There are 20 individual aminoacyl-tRNA synthetases, each synthetase being specific for each amino acid. Although the aminoacyl-tRNA synthetase is capable of covalently attaching a single amino acid to each of its degenerate tRNAs, the synthetase recognizes only one amino acid.
4. The ribosome is a macromolecular complex that catalyzes and drives the synthesis of the protein. This complex is composed to two subunits (the large 60S and the smaller 40S subunit) with each subunit being composed of both protein and ribosomal RNA (rRNA). Each subunit of the ribosome is responsible for a different function in translation: the small subunit, called the decoding center, is where the charged tRNA reads the mRNA codons, and the large subunit, called the peptidyl transferase center, where the chemical process of producing the growing peptide chain occurs.

The instructions for the correct sequence of amino acids required for the production of any protein is contained in the genetic code. The genetic code is the set of three bases, known as a codon, with each codon describing one amino acid. The codons are almost always nonoverlapping and are read as subsequent codons (i.e., base pairs are not "skipped" in the process of translation). There are 64 potential three-base combinations that comprise the genetic code. Therefore, as stated above,

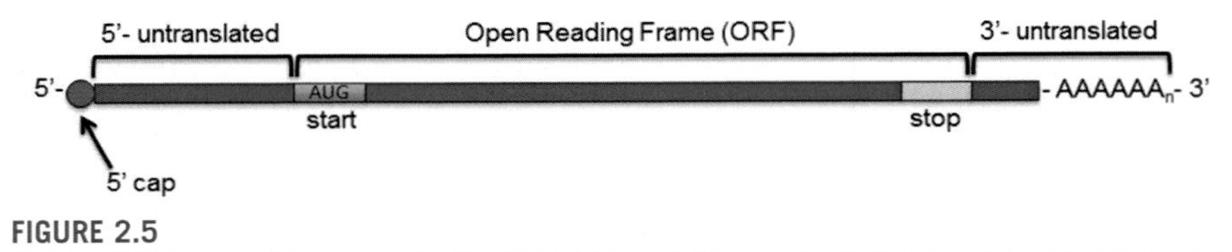

FIGURE 2.5

Illustration of a mature mRNA including the 5′-untranslated sequences, the open reading frame (ORF), and the 3′-untranslated sequences.

several different codons encode for a single amino acid. Furthermore, three of these codons (UAA, UAG, and UGA) do not encode for an amino acid. Instead, these three codons serve as signals for the termination of translation.

Translation is initiated by the formation of two independent initiation complexes. One initiation complex utilizes a series of factors to recruit the initial amino acid (almost invariantly a methionine recognizing the AUG codon) into the small 40S ribosomal subunit. The second initiation complex uses a different set of factors to recognize the 5′-cap of the mature mRNA and thereby prepare the mRNA for translation. Once these two complexes are formed, they are combined with the small ribosomal subunit; this small ribosomal complex scans along the mRNA until it encounters the initiating AUG codon, at which point the large 60S ribosomal subunit is recruited, initiating factors are released, and the complete ribosome is ready for translation.

Once initiation is complete, each subsequent charged tRNA is recruited to the ribosome through base pairing interactions between the anticodon loop of the tRNA and the codon present on the mRNA. The ribosome catalyzes the covalent attachment of the growing peptide chain onto the newly recruited amino acid, at which point a variety of factors facilitate the movement or translocation of the noncharged tRNA and the tRNA containing the growing peptide chain within the ribosome, facilitating the recruitment of the next charged tRNA and amino acid. This process continues until the termination codons are encountered, at which point a series of release factors promote the release of the completed peptide chain and the dissociation of the two ribosomal subunits.

Regulation of gene expression

All of the molecular mechanisms described to this point are necessary for the basic expression of every single gene, which is known as basal gene expression. However, human beings are extremely complex biological machines and require more than simply the basal level of expression. They must be able to turn genes on or off in response to nutrient needs, extracellular signals such as hormones and steroids, organ- or tissue-specific expression, cell type—specific expression, and temporal restrictions such as developmental stage, differentiation stage, and even the different stages of the cell cycle. To do this, the human body, and all biological species, has developed extensive and complex regulatory mechanisms to ensure that the correct genes get turned on or off at the proper time and to the proper extent.

Pretranscriptional gene regulation

The most heavily regulated step in gene expression is pretranscriptional and most often affects the process of transcription initiation. Pretranscriptional regulation utilizes regulatory sequences, which are the stretch of DNA that encompasses the complete collection of DNA elements that contribute to altering the expression of any one gene. Regulatory sequences include any or all of the following:

1. The core promoter—the minimal DNA sequence located immediately adjacent to the transcriptional start site that is sufficient to initiate transcription, usually between 60 and 120 bases in length.
2. The proximal promoter—the DNA sequence that contains the core promoter but also contains primary regulatory elements, usually up to several hundred bases in length initiating at the transcriptional start site.
3. The distal promoter—DNA regulatory sequences that are located several hundred, several thousand, or even many kilobases distant from the transcriptional start site. These elements include enhancers, silencers, or insulators.

In addition to these DNA regulatory sequences, also known as cis-acting factors, regulation requires a host of trans-acting factors. These trans-acting factors are most often proteins known as specific transcription factors, which recognize very specific sequences of DNA located within the regulatory sequences. Binding of the specific transcription factor can either promote or inhibit the expression of a gene, depending on the promoter context in which the transcription factor binds. These factors may have many different functions, which include the ability to recruit or inhibit the binding of general transcription factors to the core promoter element or promote the bending of the DNA to facilitate the interaction of additional proteins that subsequently enhance or repress gene expression.

Furthermore, higher eukaryotes tightly package their DNA into nucleosome structures to fit nearly 3 billion bases into the nucleus. A result of this packaging is that DNA regulatory elements may be inaccessible to the binding of regulatory proteins. Therefore, another function of specific transcription factors is to promote the modification or remodeling of nucleosomes to make DNA regulatory sequences more accessible to contribute to the activation of gene expression or less accessible to inhibit gene expression.

Because of the complexity of higher eukaryotes such as humans, gene regulation is very rarely, if ever controlled by a single regulatory factor. Frequently alterations in gene expression will result from a series of actions of individual regulatory factors, all of which ensure the accurate expression of a specific gene at a specific time or place. For example, consider the situation where the expression of a gene needs to be activated; however, the core promoter element is tightly compacted into nucleosomes. In this case, an initial DNA-binding protein will bind to a nearby accessible DNA regulatory sequence and recruit a histone-modifying enzyme, which covalently alters the core histones. This event relaxes the compaction of the DNA to allow a second DNA-binding protein to bind, which in turn recruits a nucleosome remodeling complex to further locally decompact the nucleosome structures. A DNA-binding protein then binds to a specific DNA sequence present in the proximal promoter to recruit the general transcription factors and RNA polymerase to the core promoter to initiate transcription (Fig. 2.6).

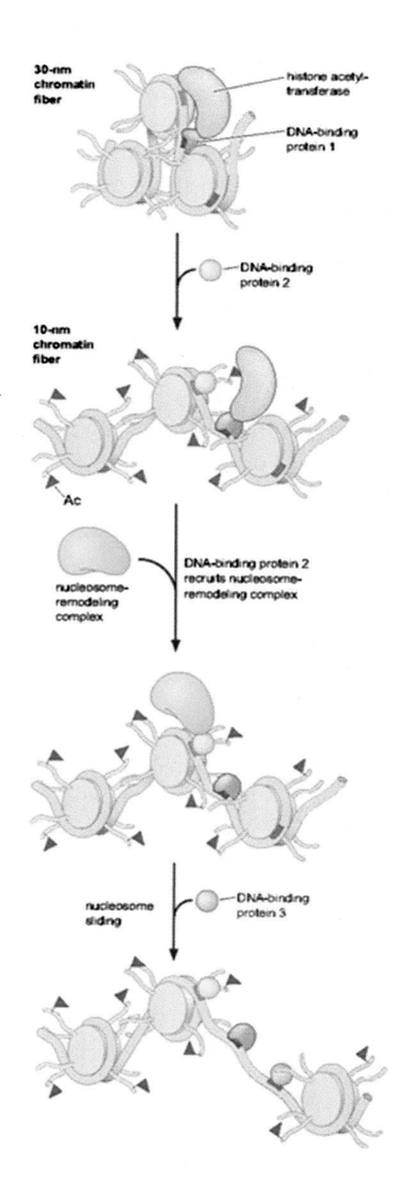

FIGURE 2.6

Model illustrating the sequential action of multiple regulatory proteins for the activation of the expression of a gene.

In addition to the binding of specific transcription factors to regulatory sequences, alterations in gene expression may occur through the chemical modification of DNA present in these regulatory sequences. The enzyme DNA methyl transferase (DNMT) is capable of covalently adding a methyl group to cytosine bases in regions of DNA rich in cytosine and guanines, known as CpG islands. The addition of the methyl group alters the weak binding interactions that are essential for promoting the interaction between a DNA-binding protein and its DNA recognition sequence, thereby inhibiting the binding of these proteins and subsequently the expression of the gene. In essence, the methyl groups create "spikes" on the DNA that prevent regulatory proteins from binding, similar to the spikes above many downtown city doors that prevent pigeons and other birds from sitting in these places.

Posttranscriptional regulation

Although transcription initiation is the most heavily regulated step in gene expression, fine-tuned regulation also occurs posttranscriptionally, most commonly through the action of microRNA (miRNA). MicroRNAs are small pieces of RNA ($\approx$24 bases) that are produced from precursor RNAs contained within transcribed mRNA (Fig. 2.7A). The transcribed mRNA will form secondary structures known as hairpin loops that are specifically recognized by the enzyme Drosha, which cleaves the hairpin from the larger RNA species. These hairpin loops can form in any part of the nonprocessed RNA, including exons, introns, and 3′- or 5′-untranslated regions (Fig. 2.7B). The enzyme dicer then further processes the released

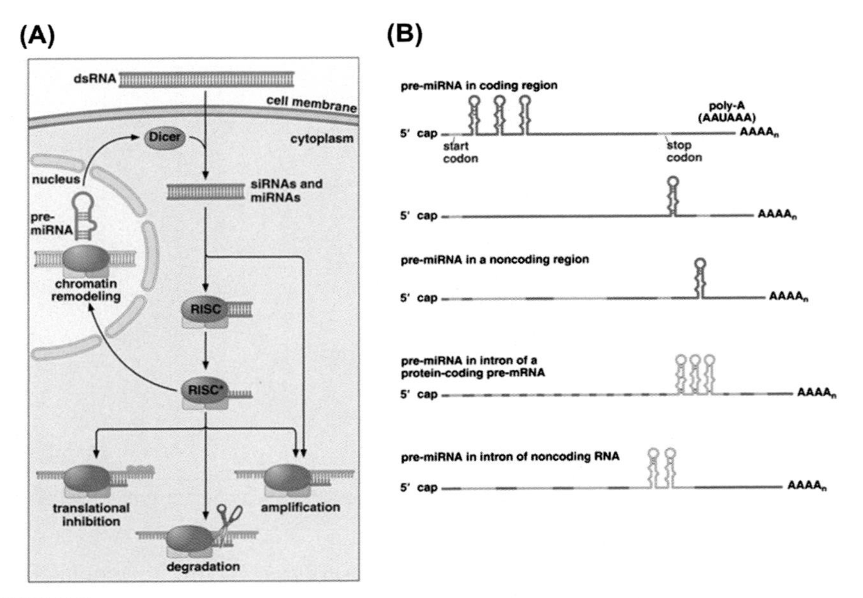

FIGURE 2.7

(A) Illustration of the processing and maturation of miRNA and (B) a schematic demonstrating the potential location of miRNA sequences in pre-mRNA species.

hairpin generating a double-stranded RNA of approximately 24 bases. This short double-stranded RNA is incorporated into the RNA-induced silencing complex (RISC), which promotes the denaturing of the double-stranded RNA, expulsion of one strand that is ultimately degraded, and retaining the other strand to serve as the guide RNA.

The guide RNA assists in directing the RISC complex to the 3′-end of the target RNA through complementarity between the guide RNA and short sequences in the 3′-end of the target RNA. The level of complementarity between the guide RNA and its target will determine the fate of the target RNA. If there is a high complementarity, the mRNA target is degraded. If there is a lower complementarity, meaning there are several bases mismatch between the mRNA and its target, translation will be inhibited. Regardless of the mechanisms, the result will be the inhibition of translation of the target RNA. Because miRNA can inhibit translation regardless of its level of complementarity, each miRNA is capable of acting on dozens or more individual genes. Therefore, changes in the expression or production of a single miRNA may have broad ranging consequences on the posttranscriptional regulation of the expression of genes. Furthermore, because miRNAs derive from transcribed genes, their expression is subject to the same transcriptional regulation described above.

The role of mutations in altering gene expression and protein function

Mutations in DNA sequences can have serious and broad ranging effects on gene expression and protein function. These mutations can involve changing of a single base, insertion or deletion of a single nucleotide, several nucleotides, or even amplifications or deletions of large regions of chromosomes. These changes will alter protein expression or function depending on where they occur.

1. Changing of a single base to a different nucleotide in a coding sequence will result in a missense mutation. In this case, the codon is altered causing an incorrect amino acid to be incorporated. The single base change may also create a nonsense mutation, which results in an amino acid codon being converted to a stop codon, creating early truncation of the protein. Finally, a single base insertion or deletion will result in a frame shift mutation, in which the reading frame is shifted, producing an incorrect protein product.
2. Mutations that occur in regulatory sequences, regardless of where these sequences are located, have the potential of altering the expression of a gene. This alteration may occur by inhibiting the ability of a transcription factor to bind to the regulatory sequence or by inhibiting the ability of a histone-modifying enzyme or nucleosome-remodeling complex to condense or decondense chromatin structures. These mutations can also affect the ability of DNA methylase to act on DNA, thereby affecting the methylation status of promoters.

3. Mutations may occur in splice site recognition sequences or in the regulatory sequences that control mRNA splicing. These types of mutations will result in improperly spliced mRNA that creates mRNA products that either have incorrectly removed an exon or retained an intron. These improperly spliced products may then generate frame shift mutations, ultimately creating a nonfunctional protein product.
4. Mutations may be present in miRNA or in the miRNA target sequence. However, because this type of regulation can accommodate a wide variability in complementarity between miRNA and its target sequence, deletion of target sequences or changes in expression levels of the miRNA will usually have the largest effects in the ability of the miRNA to regulate its target gene.
5. Finally, there can be amplification or deletion of large regions of chromosomes or gross chromosomal structural alterations. These changes will result in the overproduction of genes (amplification), complete loss of an allele of a gene (deletion), generation of disease causing fusion proteins, or placement of tightly regulated genes under the control of constitutively active promoters, the last of these two resulting from gross chromosomal structural alterations.

It is important to remember that these individual mutational events do not happen in a vacuum. The presence of a single mutation has the potential for globally changing cellular events. For example, a mutation in one transcription factor (TF1) has the potential for altering its ability to bind to regulatory sequences and directly affecting the expression of hundreds of genes. Many of the hundreds of genes that are regulated by TF1 may also be transcription factors (e.g., TF2) or regulatory proteins that then regulate the expression of even more genes. Therefore, a mutation that decreases the activity of TF1 will cause a reduction in the expression of its target gene, TF2. The reduction in TF2 will subsequently cause a reduction in the expression of its target genes. As a result, the presence of a single mutation has the potential to generate a global domino effect culminating in the alteration of the expression of hundreds if not thousands of downstream genes. Alternatively, some of these genes may be enzymes or proteins that are essential for the normal functioning of the cell. Therefore, a single mutation that inhibits the ability of a transcription factor to bind to DNA has the potential for directly or indirectly changing the expression of hundreds, if not thousands of genes.

Mutations, gene expression, and precision medicine

Understanding how mutations alter the expression of genes and potentially affect the biological or biochemical action of enzymes is key to the basis of precision medicine. Precision medicine is the ability of a clinician to select precise treatments based on the genetic underpinnings of a patient's disease. A central example in how precision medicine may influence the manner in which a clinician treats a patient is through understanding how mutations of the CYP enzymes affect their activity. CYP enzymes are proteins in the cytochrome P450 family of enzymes that serve as the major metabolizers of drugs in the liver. Mutations in different CYP family members will directly affect their enzymatic activity and subsequently their ability to metabolize drugs. Knowing how these mutations affect CYP enzymes is essential for the prescribing of drugs.

Of the CYP family of enzymes CYP2D6 is the major metabolizer of many drugs. CYP2D6 also has a well-studied genotypic and phenotypic variability due to the presence of natural allelic variants, or mutations present in the general population. These different variants have the potential of creating "poor metabolizers" by abrogating CYP2D6 activity or creating "intermediate metabolizers" by reducing its activity. Furthermore, some individuals have multiple copies of CYP2D6, resulting in an overproduction of the enzyme and causing these individuals to be "ultrarapid metabolizers."

The ability of a patient to respond to treatment depends on which natural variant of CYP2D6 they have, the drug they are receiving, and how the variant affects the metabolism of that drug. For example, a patient may be treated with a drug for which the drug must first be metabolized to its active form by CYP2D6 to be effective. A patient that has poor or intermediate metabolizing variant of CYP2D6 will be unable to generate the active metabolite, and therefore, this treatment would be ineffective. In contrast, some drugs are cleared from the system by first being metabolized to an inactive form by CYP2D6. An individual who is an ultrarapid metabolizer would therefore clear the drug too quickly to be therapeutically effective at normally given doses. In both of these cases, it is essential for the clinician to understand which genetic variant a patient has and how this variant affects the metabolism of a specific drug to prescribe a precise treatment (see Chapter 6).

Having knowledge of underlying genetic alterations is not just applicable to the ability of an individual to metabolize drugs but can also be used to understand how gene mutations and altered gene expression contribute to differences in how a patient responds to therapies. Consider the case where two male patients of similar age present to their oncologist with pathologically and histologically indistinguishable tumors. The oncologist prescribes the same course of chemotherapeutic treatment, for which one patient has a positive response, while the second patient does not. This outcome may be disappointing from a clinical perspective. However, when viewed from a molecular genetic perspective, these outcomes may make sense, since the apparent histologic and pathologic similarity between these tumors masks the underlying genetic differences that most likely exist between these patients.

The development of a tumor results from the acquisition of a series of mutations. As described above, individual mutations have the ability to alter the expression of hundreds, if not thousands of genes that then interact through complex networks of regulation. Therefore, although the two patient tumors are histologically and pathologically similar, each tumor most likely derived from a different series of mutations, which would then alter global gene expression through different mechanisms. As a result, the first patient may have mutations or altered gene expression patterns that are effectively targeted by the therapy, whereas the second patient does not contain these genetic changes and therefore will not respond as well.

In addition to directing which drugs are being prescribed and providing the potential for predicting how a patient will respond to therapy, the molecular genetic understanding of how mutations affect gene regulation has other benefits toward human health. Knowledge of causative genetic changes and how these changes affect global gene regulatory networks has the potential to determine the genetic mechanism that contributes to a particular disease. These mechanisms may then be used to direct genetic counselors who work with patients and their families, allowing them to inform the individuals of predispositions they may have to particular health issues, specific health risks they may have, and the potential of passing these genetic changes onto their children. Finally, by understanding how mutations affect global gene regulation, basic scientists have the potential of identifying novel biological targets that may be used to develop new pharmaceutical treatments for diseases that have higher specificity and fewer side effects.

CHAPTER

Fundamentals of epigenetics

3

Judy S. Crabtree, PhD [1,2]

Associate Professor, Department of Genetics, Louisiana State University Health Science Center, New Orleans, LA, United States[1]*; Scientific and Education Director, Precision Medicine Program, Director, School of Medicine Genomics Core, Louisiana State University Health Sciences Center, New Orleans, LA, United States*[2]

Chapter outline

Introduction

Chapter 2 described the central dogma of molecular biology and explained the mechanisms by which genes are expressed, much of which is dictated by the sequence of DNA in regulatory regions and within the genes themselves. Epigenetics, or an alteration that results in heritable changes that are not due to changes in DNA sequence, is another mechanism by which the cell can regulate the expression of genes in addition to classical transcriptional and translational regulation. As we are still learning, epigenetic regulation can take many forms, including histone modifications such as methylation, acetylation, and phosphorylation; DNA modifications by DNA methylation; and the actions of noncoding RNAs and microRNAs. Epigenetic regulation has important effects on embryogenesis, development, aging, genetic syndromes, and many diseases including cancer. Indeed, DNA methylation and histone modifications have been proposed for use as biomarkers for cardiovascular indications, cancer detection, and tumor prognosis and as a predictor of therapeutic responses.

Clinical Precision Medicine. https://doi.org/10.1016/B978-0-12-819834-6.00003-3

Histone modifications and variants

Chromosomes are the result of tight packaging of DNA around proteins called histones (see Chapter 1). This tightly condensed DNA is called chromatin, and one type of epigenetic regulation is based on the simple concept that chromatin status and histone modification can regulate gene expression. In general, tightly bound DNA is inaccessible to transcription factors and is silenced, whereas loosely coiled DNA is more accessible to regulatory factors and actively transcribed. Therefore, gene expression is dependent not only on the DNA sequence but also on the chromatin structure, how tightly the DNA is packaged within these structures, and the accessibility of gene regulatory regions.

A nucleosome is made up of 146bp of DNA that is wrapped around a core of eight histone proteins that contains two of each histone H2A, H2B, H3, and H4. The structure of each of these histones includes an abundance of positively charged amino acids (which attracts negatively charged DNA) and protruding N-terminal protein tails that facilitate packaging by interacting with neighboring nucleosomes. Each histone tail has multiple sites that can be chemically modified by methylation, acetylation, and phosphorylation, with certain amino acids in the tails being the preferred targets for modification (Fig. 3.1). For example, lysine (K) and arginine (R) residues are subject to methylation, acetylation occurs preferentially at lysine, and phosphorylation primarily occurs at serine (S) residues. The presence of these epigenetic "marks," which is also known as the histone code, alters the function of the nucleosome and ultimately determines gene expression based on packaging density of the chromatin. The presence and absence of histone marks is determined by families of enzymes in the cell called histone lysine methyltransferases (KMTs), histone lysine demethylases (KDMs), histone acetyl transferases (HATs), histone deacetyl transferases (HDACs), and protein kinases. These enzymes, particularly HDACs and kinases, have received considerable attention as drug targets and are a part of the standard of care chemotherapeutics for many forms of cancer. Whereas methylation, acetylation, and phosphorylation are the primary modifications studied thus far, additional modifications such as ubiquitination, citrullination, and sumoylation are relative newcomers to the field of histone modification and may also

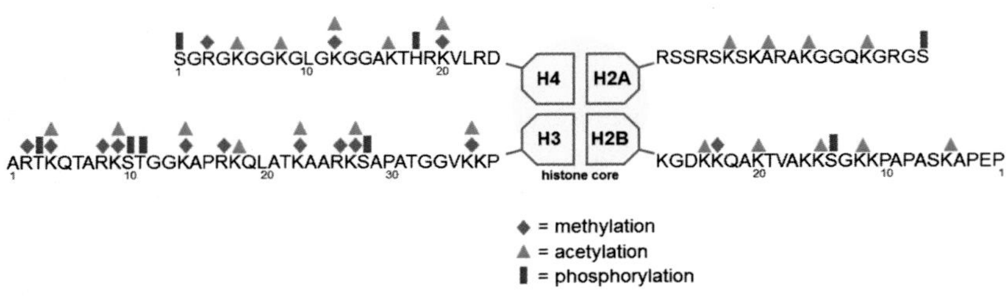

FIGURE 3.1

Histone N-terminal tail modifications.

significantly contribute to regulation by the histone code. Furthermore, whereas most of the identified and studied modifications occur on the N-terminal tails of histone proteins, modifications have also been identified in core amino acids as well as on the C-terminal ends of histones H2A and H2B.

Modified histones (i.e., histones with methylation, acetylation, or phosphorylation of tails or core amino acids) are "read" or recognized by nonhistone proteins, which then bind and recruit additional modifier proteins to the chromatin to alter the structure. For example, methylated lysines recruit proteins with chromodomains, and acetylated lysines are recognized by proteins with bromodomains. Furthermore, different histone modifications are associated with different resulting chromatin structures, and in the cases of methylation, the degree of methylation (i.e., monomethylation, dimethylation, or trimethylation) can change the impact of methylation on gene expression. Methylation of histone H3 lysine 4 (H3K4; the lysine at amino acid position 4 in histone H3) is associated with open chromatin and active transcription of genes regardless of methylation status. In contrast, monomethylation at position 9 in histone H3 (H3K9me1) is associated with gene activation, whereas trimethylation at the same position, or H3K9me3, results in tightly closed chromatin and transcriptional silencing. Despite cataloging a large number of histone modifications and having detailed knowledge of several extensively studied histone modifications, there is still much to learn about the functional role of histone modifications and the regulation of these marks in gene expression.

In addition to histone modifications by posttranslational mechanisms, there are also DNA variants present in the genes that encode histones, resulting in variant histone proteins. Variant histones called H2AX, H2A.Z, or H3.3 are derivatives of histones H2A and H3. Histone variants are produced and inserted into the nucleosome to effect a particular change in the chromatin status for a specific function. For example, histone H2AX is important in DNA repair and recombination and is inserted at the sites of double-stranded DNA breaks.

DNA methylation

Changes in chromatin structure can also arise by modification of the DNA itself. DNA methylation refers to the transfer of a methyl group from a donor molecule (typically S-adenosyl-L-methionine or SAM) to the C-5 position of a cytosine residue in DNA, resulting in 5-methylcytosine (m5C) (Fig. 3.2). Since the position of the newly added methyl group is on the outside of the DNA double helix in the major groove, the presence of m5C does not disrupt normal base pairing with its partner nucleotide, guanine.

The presence of methylated residues in DNA can directly alter gene expression by interfering with chromatin structure and the function of DNA-binding transcription factors. As a general rule, highly methylated or hypermethylated DNA sequences are typically found in regions of tightly packed chromatin and are not expressed, whereas nonmethylated or low-level DNA methylation correlates with

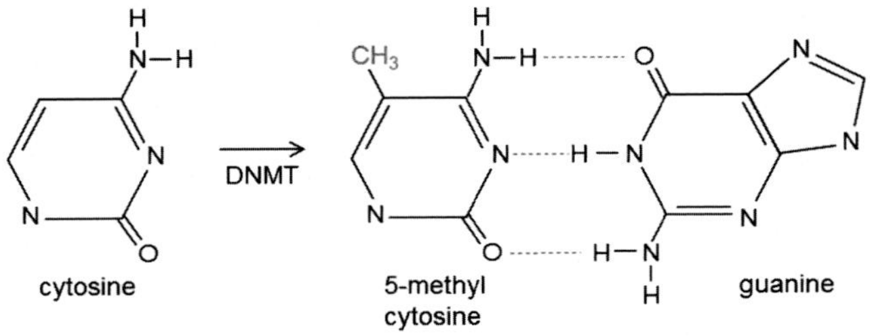

FIGURE 3.2

Cytosine methylation in DNA. DNA methyltransferases (DNMTs) add methyl groups to the 5 position in the cytosine ring. The addition of this methyl group to cytosine does not interfere with the hydrogen bond base pairing with guanine, and the methyl groups protrude into the major groove of the double helix.

loosely packed chromatin and active gene expression. In the oncology field, DNA hypermethylation is an important mechanism in the silencing of tumor suppressor genes, and DNA hypomethylation results in abnormal transcription and/or increased gene expression of oncogenes. In recent years, both hypo- and hypermethylated biomarkers have been validated for use in cancer research and diagnostics.

Methylation is also a normal part of genome maintenance, and in mammals, the majority of DNA methylation is located in regions of the DNA called CpG islands. CpG islands are typically located near gene regulatory and promoter regions, and more than 80% of the cytosines in these dinucleotide repeat regions are methylated. Enzymes called DNA methyl transferases (DNMTs) perform the addition of methyl groups to DNA, and in humans, there are three major DNMTs—DNMT1, DNMT3A, and DNMT3B. DNMT1 is the enzyme responsible for maintenance of methylation marks as cells divide. During replication, the new strands of copied DNA are hemimethylated for a brief period of time, meaning only the parental strand is methylated, until DNMT1 adds methyl groups in the new chains at the same locations as in the parental strand. This mechanism ensures that DNA methylation patterns are preserved from parental to daughter cell throughout cell division.

The methylation marks present in DNA also change during early development and gametogenesis (Fig. 3.3). In gametogenesis, primordial germ cells are methylated until they enter the genital ridge. Once in the genital ridge, all parental epigenetic marks are erased; then as the primordial germ cells mature and differentiate into sperm or eggs, de novo methylation patterns and imprinting marks are newly established. Once sperm and egg unite in a fertilized oocyte, there is global demethylation in the early embryo that continues until the early blastula stage. As cells mature into different cell lineages, DNMT3A and DNMT3B are responsible for establishing de novo methylation patterns that persist in the new organism.

In mammals, the DNMT enzymes are critical to embryogenesis and survival. Mice lacking Dnmt1 or Dnmt3b exhibit embryonic lethality and genome-wide loss of methylation. Mice lacking Dnmt3a die in the early postnatal period and

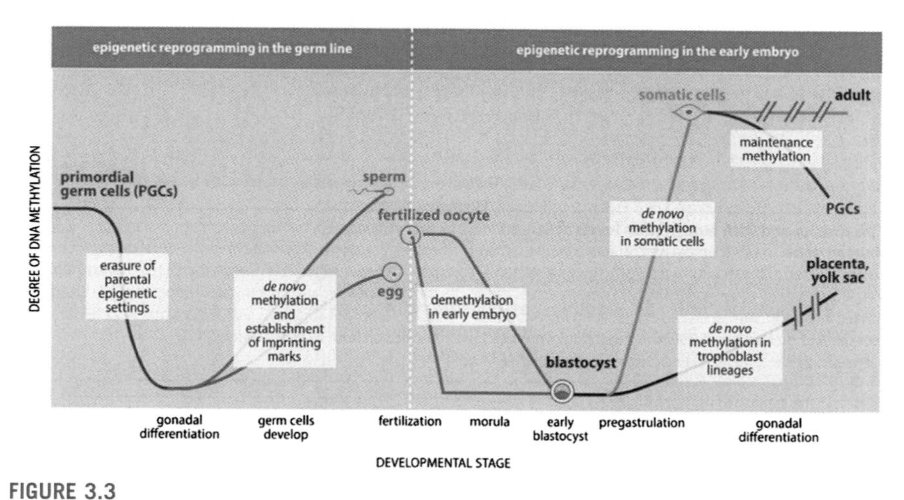

FIGURE 3.3

DNA methylation in development.

exhibit failure to establish methylation imprints in germ cells. In humans, loss of DNMT3B results in a clinical syndrome called ICF syndrome, so named for the phenotype present in these individuals of immunodeficiency, centromeric DNA instability, and facial anomalies.

Genomic imprinting

Under normal circumstances, we inherit two copies of every chromosome and gene—one copy from our mother and one from our father. For some genes, both copies of each gene are functional (termed biallelic expression), and it does not matter which allele is maternally or paternally derived. For other genes, only one allele is expressed (monoallelic expression), and the inactivated allele is randomly silenced with no regard to whether the allele was inherited maternally or paternally. Finally, genes with monoallelic expression can have consistent inactivation of either the maternal or the paternal allele, even though the DNA sequence of these two alleles is identical. This is called DNA imprinting and is the direct result of epigenetic patterning between the two alleles.

Prader-Willi/Angelman syndrome region

Some gene regions of the genome are regulated by a combination of epigenetic mechanisms that imprint either the maternal or the paternal alleles. The Prader-Willi/Angelman syndrome (PWS/AS) region on human chromosome 15 is a classic example of genomic imprinting and the manifestation of a genetic disorder. Loss of a functional *paternal* copy of 15q11-q13 results in PWS, whereas loss of a functional *maternal* copy of UBE3A, a gene within 15q11-q13, causes AS. This occurs because the PWS/AS region of chromosome 15 contains a number of genes that are regulated by the PWS/AS imprinting center (PWS/AS IC). This region of chromosome 15 is imprinted in all individuals such that the maternal alleles are silenced and the paternal alleles are expressed. The parent-specific imprinting occurs by epigenetic modifications including DNA methylation at CpG sites as well as histone modifications including H3K4me, H3K9me, H3 and H4 acetylation, and H4K20me (Fig. 3.4). In this gene region, the maternal alleles for *MKRN3, NDN,* and *SNRPN* are silenced by maternal-specific methylation of CpGs located in the promoters or exon 1 of these genes. There is also parent-specific histone modification, with the *NDN* gene exhibiting paternal-specific H3K4 methylation. The *SNRPN* gene has histone acetylation and H3K4 methylation occurring only on the paternal allele, while there is maternal-specific methylation of H2K9 and H4K20. In syndromic individuals, the genetic loss of the imprinted paternal allele by gene deletion, uniparental disomy (where the both copies of a chromosome come from the same parent), sporadic mutation, or chromosomal translocation results in PWS because the remaining maternal allele is epigenetically silenced. Conversely, deletion of the same region on the maternal chromosome causes AS. More details and information on the genetics of PWS and AS are available from recent reviews.[1,2]

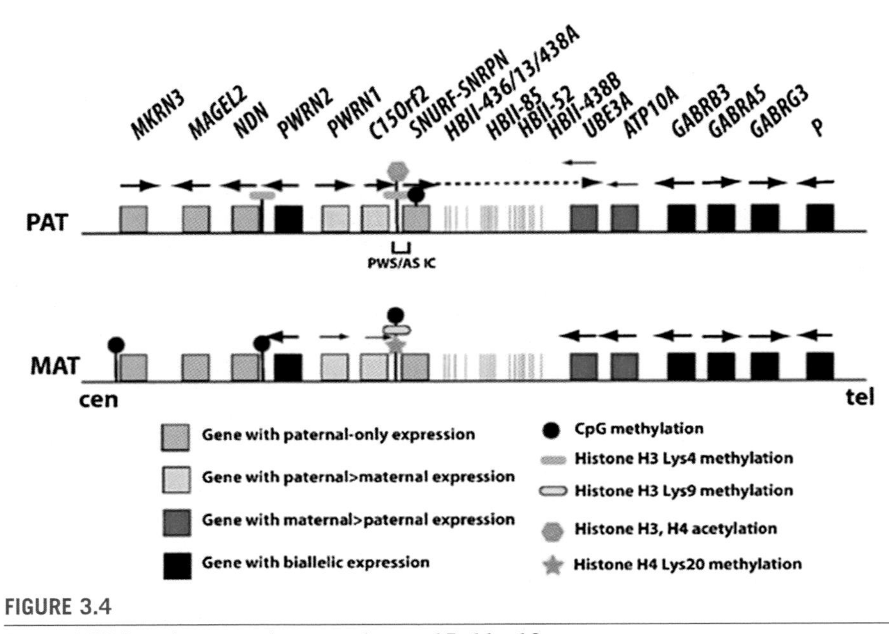

FIGURE 3.4

Prader-Willi/Angelman syndrome region on 15q11-q13.

Noncoding RNAs and microRNAs

In addition to encoding genes and regulatory sequences, the human genome also includes many gene regions that do not encode for proteins, but which have direct impacts on gene expression. These gene regions encode long noncoding RNAs (lncRNA; >200 nucleotides in length) and small microRNAs (miRNA; 19–22 nucleotides) that play regulatory roles in the cell and influence the expression of protein-coding genes. Mammalian lncRNAs are important for both genetic and epigenetic regulation of gene expression because they are often tissue specific and function by recruiting chromatin-modifying machinery to specific locations within the genome. In many cases, lncRNA acts as a tether that links proteins/protein complexes to particular regions in the DNA through nucleotide base pairing between the lncRNA and the DNA. Other times, this process happens without the use of nucleotide base pairing to DNA. Instead, the lncRNA folds itself into secondary and tertiary structures that are specifically recognized by chromatin-modifying protein complexes and facilitate recruitment to the chromatin. Finally, lncRNAs can function as coregulators of gene expression by blocking or recruiting activators and/or repressors to protein-coding genes.

MiRNAs are small RNAs that function to block gene expression in the cell by base pairing to complementary target mRNA and either stimulating degradation of the mRNA or blocking translation of the mRNA into protein. MiRNAs are produced from an independent gene that makes a primary RNA (pri-miRNA) or from an intronic piece of RNA that was spliced from a host gene, often called a mirtron (Fig. 3.5). These initial long RNA species are acted upon by enzymes that generate a pre-miRNA, which has a hairpin stem-loop structure. Pre-miRNAs are exported out of the nucleus and are acted upon by a protein called dicer, which removes the loop at the end to create a 19–22 bp double-stranded duplex of RNA. This RNA is then incorporated into the RNA-induced silencing complex and used as a guide RNA to find complementary sequences in the mRNAs present in the cytoplasm. MiRNA-binding sites in mRNAs are predominantly found in the 3′UTR of the target mRNA. MiRNAs have a "seed" sequence that is nucleotides 2–8 of the miRNA, and for the miRNA to recognize an mRNA as a target, the seed sequence must have perfect complementarity. The degree of complementarity of the other nucleotides in the miRNA determines whether the mRNA is targeted for degradation (greater degree of complementarity) or blocked from being translated into protein (lower degree of complementarity). Just as with variants in protein-coding genes, variants in miRNA that alter expression of miRNAs and/or complementarity of miRNAs for their mRNA targets have been implicated in many disease states, and miRNA-based therapeutics are in development.

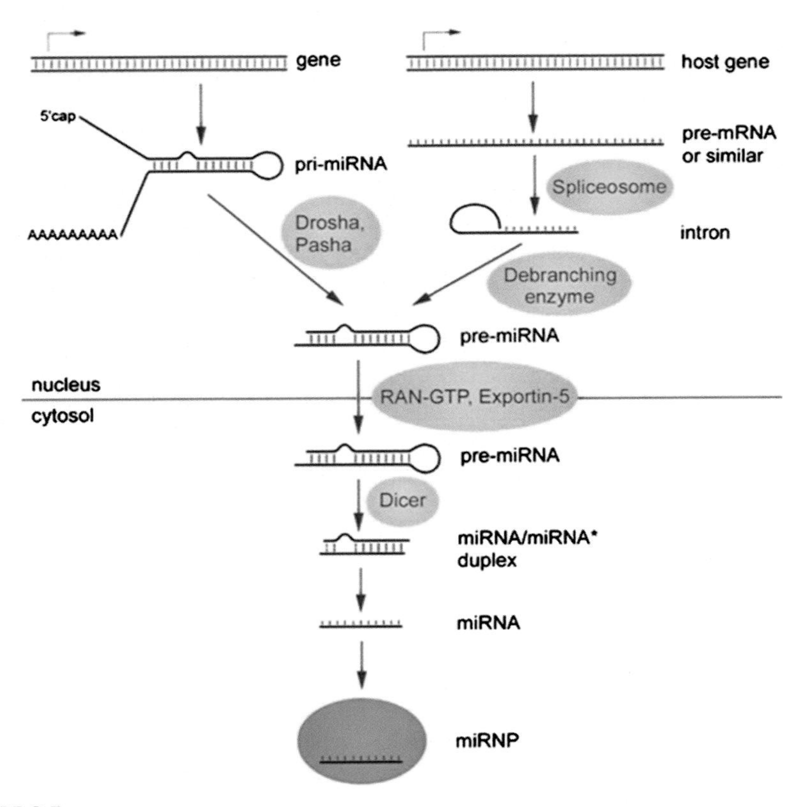

FIGURE 3.5

MicroRNA biogenesis.

X chromosome inactivation

Females have two X chromosomes, and males have one X and one Y chromosome. In females, the expression of genes from both X chromosomes would result in a potentially harmful double dose of X chromosome genes. To control for this dosage imbalance between males and females, females have a mechanism called X chromosome inactivation to ensure that in females, any extra X chromosomes are silenced, resulting in equivalent expression of X chromosome genes in both males and females. Furthermore, X chromosome inactivation provided the first early evidence that modifications in DNA could alter gene expression.

The decision as to which X chromosome is inactivated (maternal or paternal) is random and occurs in the progenitor sex cells during early embryogenesis. Either the maternal or the paternal X is selected for inactivation, and that choice is maintained through all subsequent generations of daughter cells. Because of this random inactivation, females are a mosaic of cells in which either the maternal or paternal X is silenced. A perfect example of this mosaicism is the calico cat (Fig. 3.6). In cats, the gene for fur pigmentation is located on the X chromosome, and depending on which copy of the X chromosome is inactivated (maternal or paternal), the coat color is expressed as either orange or black. Because X inactivation happens primarily in females, the majority of calico cats are female. X inactivation is stable throughout cell division, but not across generations because epigenetic marks like methylation are erased and reset during early embryogenesis. It is equally likely that a maternal X or paternal X will be inactivated in a daughter, regardless of which allele was inactivated in her mother.

X inactivation in mammals occurs through an antisense pair of noncoding RNAs expressed from the *XIST* and *TSIX* genes that are located in the X inactivation center or the XIC on the X chromosome. Before inactivation is triggered, both of these genes are weakly expressed in both male and female cells and produce a lncRNA product. Once inactivation is initiated, these genes are differentially regulated, with the *XIST* gene being expressed on the chromosome to be inactivated, and the *TSIX* gene silenced. The noncoding RNA product from the *TSIX* gene is the

FIGURE 3.6

Calico cat. Random inactivation of either the maternal or paternal X chromosome via X inactivation results in the patchy coat color of calico cats. Ksmith4f via Wikimedia Commons.

antisense of XIST and functions as an inhibitor of XIST. Therefore, when TSIX is present, XIST is inhibited. In the absence of TSIX, the XIST lncRNA coats the chromosome to be inactivated, spreading from the XIC and marking the chromosome as silenced. Rep A is another lncRNA that inhibits TSIX, further amplifying the function of XIST. Rep A also promotes methylation of the *TSIX* gene region by recruiting a protein called PRC2, which further inactivates the selected X chromosome.

The great majority of genes on the inactivated X chromosome are silenced. The DNA in the inactivated chromosome is highly methylated compared with the active X chromosome and is packaged into tightly condensed heterochromatin called a Barr body. The inactivated chromosome also has modified histones with low levels of generalized histone acetylation, low levels of H3K4 methylation, and high levels of H3K9 and H3K27 methylation. The histone methylation marks are placed by PRC2 (and its associated proteins) to regulate the chromatin condensation and facilitate the production of Barr bodies.

While most of the genes on the silenced X chromosome are not expressed, up to a quarter of the genes are expressed. The *XIST* gene, for example, is always expressed at a high level on the inactivated chromosome. Other genes that escape silencing are typically genes that are also present on the Y chromosome in regions that are called pseudoautosomal. This is because, like the autosomal chromosomes, these gene regions require expression from two alleles to achieve the appropriate gene dosage, unlike the genes that are specific to a sex chromosome. No dosage regulation is required for these genes in females, so these regions escape X inactivation, although the specific mechanisms are unclear.

Conclusions

Epigenetics, or the modulation of gene expression by mechanisms other than changes in DNA sequence, is a key player in understanding how gene expression is controlled. The functions of the epigenome play a key role in regulating temporal expression of genes by modulating transcription factor binding and chromatin structure. The impact of DNA methylation, histone modification, and lncRNAs has been correlated with gene expression and chromatin structure through international programs such as the ENCODE (Encyclopedia of DNA elements) Consortium (www.encodeproject.org). The influence of lncRNAs and miRNAs is beginning to become clearer with the ever-increasing volume of literature to support their previously unknown functions. As precision medicine continues to mature, epigenetic mechanisms will further reveal their role in disease pathology, particularly in cancer where epigenetic variants such as alterations in methylation signatures are fast becoming more frequent than genetics variants.

References

1. Angulo MA, Butler MG, Cataletto ME. Prader-Willi syndrome: a review of clinical, genetic, and endocrine findings. *J Endocrinol Investig* 2015;**38**(12):1249–63.
2. Horsthemke B, Wagstaff J. Mechanisms of imprinting of the Prader-Willi/Angelman region. *Am J Med Genet* 2008;**146A**(16):2041–52.

CHAPTER

Fundamentals of heredity

4

Judy S. Crabtree, PhD [1,2]

Associate Professor, Department of Genetics, Louisiana State University Health Science Center, New Orleans, LA, United States[1]*; Scientific and Education Director, Precision Medicine Program, Director, School of Medicine Genomics Core, Louisiana State University Health Sciences Center, New Orleans, LA, United States*[2]

Chapter outline

Introduction

For centuries, many explanations were put forth to explain the transmission of physical characteristics from parent to offspring. It was not until the mid-19th century (1865) when Gregor Mendel focused his study on easily characterized traits of the garden pea plant (*Pisum sativum*) that the link was defined between physical characteristics (phenotype) and genetic makeup (genotype). In his studies, Mendel proposed that "particulate factors" were the carriers of heredity, and in future studies by Fred Griffith[1] and others[2] in the 1900s, it was discovered that the "particulate factor" he proposed was deoxyribonucleic acid or DNA. These studies ultimately led to the fundamentals of heredity in use today that we now call "Mendelian" genetics.

The human genome is the entire genetic makeup of a human and is comprised by the DNA found within the nuclear chromosomes and the mitochondria. The mitochondrial genome is a double stranded, circular molecule that contains 16.5 kilobases (kb or 16,500 base pairs), whereas the DNA found within the nucleus is present in 23 pairs of chromosomes (Fig. 4.1) for a total of roughly 3 billion bases. Chromosomes 1 through 22 are autosomes and range in length from 250 megabases

Clinical Precision Medicine. https://doi.org/10.1016/B978-0-12-819834-6.00004-5

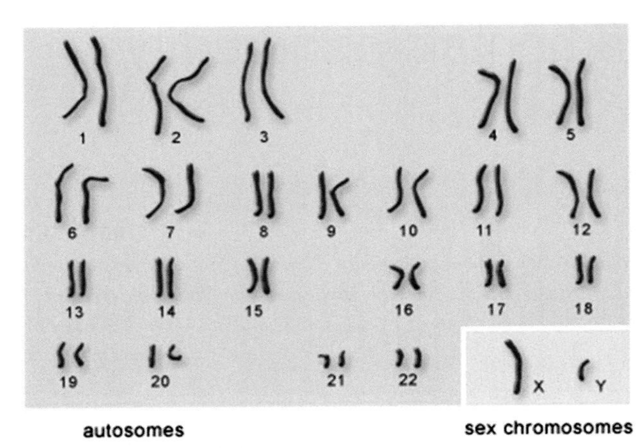

FIGURE 4.1

Human chromosomes.

Credit: US National Library of Medicine.

(Mb) to just under 50 Mb. The sex chromosomes are X and Y, with females carrying two copies of the X chromosome and males with one copy each of X and Y. Chromosomes contain not only genes that encode for functional proteins but also repetitive elements, regulatory regions, and hypervariable elements. The chromosome pairs present in offspring are inherited from the parents - with one chromosome inherited from each parent. The genes present on these chromosomes are called alleles (or gene loci)—the gene inherited from the mother is the maternal allele, the gene from the father is the paternal allele—and as we will see in subsequent sections, these alleles may contain variation, which is a difference in the DNA sequence. When a variant is present on only one of the two alleles, the individual is said to be heterozygous for that variant. If the variant is present on both of the alleles, the individual is homozygous for that variant. The sex chromosomes are slightly different than autosomes since most of the sequences on the X chromosome do not have an equivalent on the Y (the Y chromosomes is also much smaller). Females carry two X chromosomes and therefore can be heterozygous or homozygous for any allele on the X chromosome, but males are said to be hemizygous for these loci because they carry only one copy of X and one of Y.

Germline variants are the changes in DNA sequence that are inherited from parents. These variants can be non-disease-causing between individuals or groups of individuals (e.g., variations in eye color), or they can be pathogenic (e.g., variants in the BRCA1 or BRCA2 genes that cause breast and ovarian cancers). Pathogenic variants result in inheritance patterns that are tracked through families based on the resulting phenotype. In humans, any genetic trait or phenotype is likely the result of the expression of multiple genes along with contributions from environmental factors. However, some phenotypes are the result of variation or changes in DNA sequence at a single gene locus, meaning that a variant is required for the trait to be expressed. Although single-gene disorders are relatively rare and most common genetic disorders depend on multiple genetic loci, an understanding of fundamental inheritance patterns is critical and most easily explained using single-gene disorders.

Pedigrees

A pedigree is a graphical representation of a family tree that uses symbols to represent different individuals and assists with the recognition of genetic patterns. At the most fundamental level, pedigree symbols include a square for males, a circle for females, diamonds for individuals of unknown gender, and lines to indicate relationships. Affected individuals are indicated by filled symbols, carriers have a dot within the symbol, and a diagonal line through the symbol indicates the individual is deceased. Single horizontal lines indicate a mating pair, and double horizontal lines indicate consanguineous mating.

Autosomal dominant

Autosomal dominant inheritance means that only one of the two alleles carries a pathogenic variant (i.e., a heterozygous individual). In autosomal dominant inheritance, a pathogenic variant is inherited from either the mother or the father, and affected individuals can be male or female, with both being equally likely to transmit the disorder. Furthermore, affected individuals are found in every generation of a family (Fig. 4.2). Examples of autosomal dominance are multiple endocrine neoplasia, type I, some forms of polycystic kidney disease, and Von Hippel-Lindau syndrome.

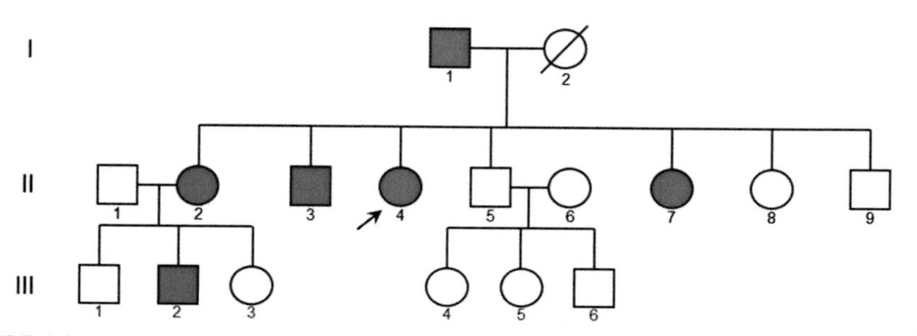

FIGURE 4.2 Autosomal dominant inheritance.

The patient (or proband) first encountered in the clinic is indicated by the arrow, patient II-4. Affected individuals are present in all three generations of this family.

Autosomal recessive

Another germline variant inheritance pattern is autosomal recessive, where two alleles of a pathogenic variant are required to cause disease—most notably in which a person receives a detrimental allele from each parent (Fig. 4.3). Males and females are affected equally, but the phenotype tends to skip a generation since an affected individual will only pass on one (affected) chromosome to their offspring, assuming their mate is not affected or a carrier. Consanguinity in families increases the risk of developing disease from autosomal recessive disorders due to increased probability of carriers within the family. Examples of autosomal recessive inheritance include cystic fibrosis, sickle cell anemia, and Tay-Sachs disease.

X-chromosome inactivation and X-linked inheritance

The presence of too many or too few chromosomes, which is known as aneuploidy, results in very severe and often lethal consequences due to changes in gene dosage. Tight regulation of genes controls the amount of gene product made and too much or too little can have harmful consequences. For example, trisomy 21 is present in a person who has three copies of chromosome 21 (Down syndrome).

The sex chromosomes are slightly different. The Y chromosome only contains around 70 protein-coding genes, many of which also have counterparts on the X chromosome or are male specific. Because females have two copies of the X chromosome, which carries more than 800 protein-coding genes, females have a mechanism called X-chromosome inactivation by which they can control gene dosage from the X chromosome. X-chromosome inactivation ensures that in females, one of the chromosomes is silenced such that males and females have equivalent expression of the genes on the X chromosome. In other words, men are *constitutively* hemizygous for the X chromosome (since they only have one), and women are *functionally* hemizygous (since one copy is silenced).

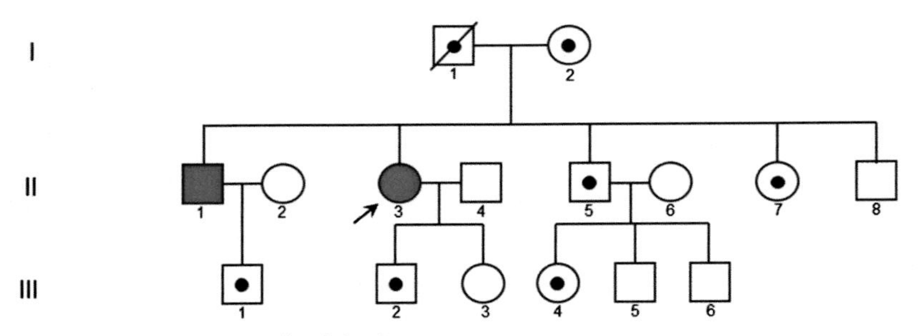

FIGURE 4.3 Autosomal recessive inheritance.

The patient (or proband) first encountered in the clinic is indicated by the arrow, patient II-3. Affected individuals are present in generation II. Note that generation III contains only carriers.

The details of X-chromosome inactivation are found in Chapter 3. Briefly, the mechanism of X-linked inactivation begins in the early embryo. If the cell detects more than one X chromosome, any and all additional X chromosomes are silenced and tightly condensed into a transcriptionally inactive form called a Barr body. The decision to inactivate an X chromosome in humans is a random process that occurs in the preimplantation embryo, around the eight-cell stage. Some cells will inactivate the paternal X chromosome, and others will inactivate the maternal X chromosome. Once the cell makes a decision as to which chromosome is inactivated, all subsequent cellular progeny will retain the same maternal- or paternal-derived X inactivation.

Because males pass only the Y chromosome to their sons, X-linked disorders have unique inheritance patterns in which there is no male-to-male transmission. Similar to autosomal inheritance, X-linked disorders can be dominant or recessive. X-linked dominant inheritance means that a single pathogenic X chromosome is sufficient to cause disease in both males and females (Fig. 4.4A). Oftentimes, this pattern includes affected females passing a pathogenic X chromosome to both sons and daughters in equal percentages. An affected son will always have an affected mother. However, affected fathers do not transmit the phenotype to their sons (since fathers contribute the Y chromosome to sons) but can pass traits onto their daughters through the pathogenic X chromosome. An affected daughter will typically have a milder phenotype as a result of X-inactivation—meaning the mutant

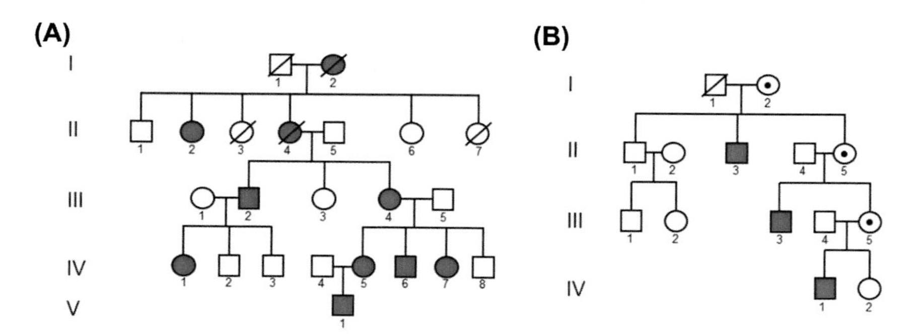

FIGURE 4.4 X-linked inheritance.

(A) X-linked dominant inheritance. Both males and females are affected, and the inheritance pattern appears to be standard dominant inheritance until it is noted that all affected males (III-2, IV-6, and V-1) have an affected mother. Furthermore, there are no affected sons resulting from affected males as indicated by III-2, IV-2, and IV-3. These observations indicate an X-linked dominant inheritance pattern. (B) X-linked recessive inheritance. Carrier females are indicated by dots within circles, and affected males are filled in boxes. Affected males II-3, III-3, and IV-1 have inherited a pathogenic X chromosome from their mothers. Note that males have a 50-50 chance of inheriting a pathogenic X chromosome from their mother—note that individual II-1 inherited a normal X chromosome from his mother, I-2.

allele is located on an inactivated X chromosome in a proportion of their cells. Some X-linked dominant diseases such as Aicardi syndrome are fatal to boys early in life, and only females with the milder phenotype are able to survive and reproduce. Examples of X-linked dominant disease include Rett syndrome, Alport syndrome, and fragile-X syndrome.

X-linked recessive disorders are more common than X-linked dominant and affect primarily males who are born to unaffected parents (Fig. 4.4B), with the mother of an affected male, typically a carrier. Males develop disorders in this case because they have only one X chromosome. Females who inherit a pathogenic X chromosome from their mothers will be carriers of the genotype but typically will not develop disease because they have a second normal X chromosome to compensate for the pathogenic one. Common examples of X-linked recessive inheritance include red/green color blindness and hemophilia A.

Y-linked inheritance

Although there are portions of the Y chromosome that are in common with the X chromosome to facilitate meiosis in males, the majority of the Y chromosome is nonrecombining and male specific. Y-linked inheritance traits and disorders are only found in males and never in females, due to transmission of a variant Y chromosome (Fig. 4.5A). The father's traits are passed to all of his sons, and dominance is irrelevant since there is only one copy of the Y-linked genes (hemizygous).

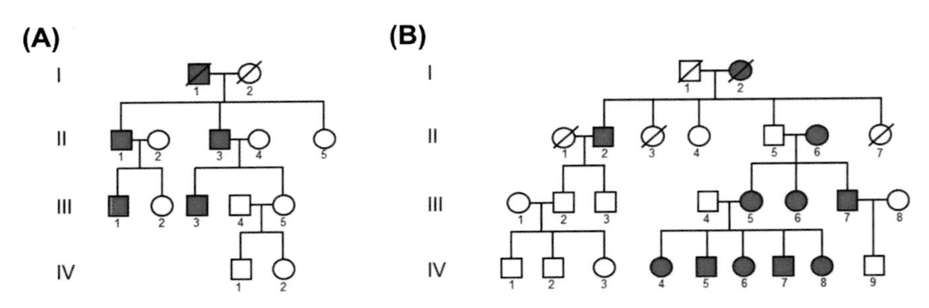

FIGURE 4.5

(A) Y-linked inheritance Males carrying a pathogenic Y chromosome will transmit to all of their sons, but none of their daughters. (B) Mitochondrial inheritance. An affected mother will pass her variant mtDNA to all of her children. Affected males do not typically pass on mitochondrial variants to their offspring as shown in the lineages from individuals 11-2 and 111-7. However, recent evidence demonstrates that in extremely rare cases paternally derived mtDNA may also be transmitted to offspring.

Mitochondrial inheritance and variable heteroplasmy

The DNA present in the mitochondrial genome is also subject to mutation and inheritability. Mitochondrial disorders tend to have highly variable phenotypes and affect tissues with a high energy requirement such as the brain and muscles. Mitochondrial disorders affect males and females although the longstanding dogma is that only females can transmit the condition to their children. Sperm does contribute mitochondrial DNA (mtDNA) to the developing zygote, but in most cases the paternal mtDNA is rapidly degraded in the developing embryo,[3] resulting in transmission exclusively via the maternal mtDNA (i.e., matrilineal inheritance; Fig. 4.5B).

Each cell contains multiple mitochondria, and there are several hundreds to thousands of copies of mtDNA per cell. Homoplasmy is when every single mtDNA carries the same pathogenic variant. The more common scenario is heteroplasmy where affected individuals have both normal and pathogenic mtDNA coexisting within cells at varying percentages. This variation in the ratio of normal to pathogenic is responsible for the vast spectrum of phenotypes associated with mitochondrial disorders. Furthermore, the inheritance of mtDNA from a heteroplastic mother to her offspring is also unpredictable, with the offspring having higher or lower percentages of pathogenic mtDNA. Recent work has demonstrated compelling evidence for paternal mtDNA inheritance as well as maternal.[4] The mechanisms by which this happens are still under investigation, and future models will likely address heteroplasmy and increased paternal mtDNA abundance in developing zygotes.

Complications to Mendelian inheritance

Real-life inheritance patterns are rarely as defined as the examples above, and many times there are complications that hinder the accurate identification of inheritance patterns. Recessive conditions can mimic a dominant pedigree pattern if a recessive trait is common in a population of people. This often occurs if an individual with a recessive condition mates with a carrier, resulting in offspring being homozygous for the recessive trait. One example of this is blood group O.

A second complication is related to the concept of penetrance. Penetrance is the probability that a person with a dominant genotype will express the associated trait. Theoretically, a dominant disease should have 100% penetrance since by definition the disease is caused by a variant in one allele (a heterozygote). However, this is not always the case. Nonpenetrance is the state in which a genetic variant is present, but the individual does not develop the disease, for example, a disease-free individual with both a parent and a child afflicted with a dominantly inherited disease. Another example is a woman with a BRCA1 mutation who never develops breast or ovarian cancer despite living a long, full life. The cause of nonpenetrance is not fully understood and is often attributed to a combination of unusual genetic background, environment, lifestyle factors, or pure chance. Genetic counselors are trained to recognize changes in penetrance for dominant conditions and advise accordingly.

Penetrance can also change with respect to the age of the individual and the predicted age of disease onset for a particular condition. The genotype of an individual confers risk for developing a disease, but the disease does not manifest until later in life. For example, Huntington disease is a progressive, age-related neurodegeneration that may appear nonpenetrant in younger individuals because they have not yet developed disease.

A third complication of Mendelian inheritance is related to the concept of penetrance and nonpenetrance and is called variable expressivity. Variable expressivity means that all the affected family members have the same genetic variant, but each member has a slightly different phenotype, which can vary from mild to severe. As with nonpenetrance, variable expressivity is probably caused by a combination of other genes involved, the environment of the individual, or lifestyle factors, many of which are not known. Recessive disorders are typically less variable than dominant disorders probably because dominant disorders require a balance between the effects of the two alleles in a heterozygote (one normal and one variant) such that the outcome is more sensitive to outside influences than a homozygote.

Finally, anticipation is the tendency of some disorders to become more severe in successive generations. This occurs most often in genetic disorders where the variant is unstable, for example, triplet repeat expansion disorders where the variant repeat gets larger and larger with each successive generation (and hence, and earlier and earlier age of onset).

Conclusion

The fundamentals of heredity are well delineated and explained for rare, single-gene disorders that include the autosomal dominant/recessive, X-linked dominant/recessive, Y-linked, and mitochondrial inheritance patterns. However, genetic disease is rarely so well defined, and complications exist in terms of penetrance and variable expressivity. Many diseases such as cancers, metabolic syndrome, diabetes, and obesity have not only multigene genetic roots but also significant contributions from the environment, which complicate the search for therapeutic interventions.

References

[1] Griffith F. The significance of pneumococcal types. *J Hyg* 1928;**27**:113–59.

[2] Avery OT, Macleod CM, McCarty M. Studies on the chemical nature of the substance inducing transformation of pneumococcal types : induction of transformation by a desoxyribonucleic acid fraction isolated from pneumococcus type iii. *J Exp Med* 1944;**79**: 137–58.

[3] Zhou Q, Li H, Li H, Nakagawa A, Lin JL, Lee ES, Harry BL, Skeen-Gaar RR, Suehiro Y, William D, Mitani S, Yuan HS, Kang BH, Xue D. Mitochondrial endonuclease G mediates breakdown of paternal mitochondria upon fertilization. *Science* 2016;**353**:394–9.

[4] Luo S, Valencia CA, Zhang J, Lee NC, Slone J, Gui B, Wang X, Li Z, Dell S, Brown J, Chen SM, Chien YH, Hwu WL, Fan PC, Wong LJ, Atwal PS, Huang T. Biparental inheritance of mitochondrial DNA in humans. *Proc Natl Acad Sci USA* 2018;**115**: 13039–44.

CHAPTER

Clinical genetics

5

Stephanie Kramer, MS, CGC

Certified Genetic Counselor, Center for Advanced Fetal Care, Clinical Assistant Professor, University of Kansas Health System, Women's Specialties Clinic, Kansas City, MO, United States

Chapter outline

In the clinic, one of the first known uses of precision medicine could be the evaluation of family medical history information. Practitioners have used family history information to help guide patient care for decades. If a patient has a significant family history of diabetes, screening for diabetes may be recommended at an earlier age than typical. When a patient has multiple relatives who have hypertension, more frequent blood pressure screening may be recommended. Clinicians have long recognized that many of these multifactorial diseases have a genetic component to them and modify care based on this premise. The medical community is now beginning to realize the potential of using genetic testing information to provide individualized specialized care to patients.

Precision medicine is expected to become even more mainstream, as time goes on and the number of testing options increases. Precision medicine has the ability to increase the efficiency and specificity of treatments and surveillance strategies for patients.

Prenatal genetic testing/screening

Evaluation of a pregnant patient's family and personal medical histories is a very important factor in determining appropriate testing and screening options. These options include invasive prenatal testing and screening, carrier screening, and clinical screening for disease.

Pregnant women who are at increased risk to have a baby with a chromosome anomaly, such as Down syndrome (trisomy 21), are offered the option of invasive

Clinical Precision Medicine. https://doi.org/10.1016/B978-0-12-819834-6.00005-7

prenatal testing, such as amniocentesis or chorionic villus sampling. Amniocentesis, which is typically performed after 16 weeks of gestation, is a procedure in which amniotic fluid is withdrawn from around the fetus. Fetal cells are extracted from this fluid and are grown in culture. Chorionic villus sampling, which is performed between 10 and 14 weeks gestation, involves obtaining a biopsy of the chorionic villi within the placenta. These placental cells are then grown in culture. Chromosome analysis and other genetic testing can be performed on cultured amniotic fluid or chorionic villi cells. Both procedures carry a risk for miscarriage of between 1/200 and 1/1000.

Testing options available on specimens from invasive prenatal procedures are varied and are dependent on the patient's personal risks and history. Available testing includes preliminary fluorescence in situ hybridization (FISH) study for a rapid evaluation of the number of chromosomes 13, 18, 21, X, and Y. FISH is a cytogenetic technique in which fluorescent probes are annealed to certain DNA sequences on a chromosome to be able to see the presence or absence of the specific area of interest, which can be an entire chromosome or part of a chromosome. FISH can also be used to evaluate for certain microdeletion or duplication syndromes.

Chromosome analysis of prenatal specimens has been available since the late 1960s. A newer technology, called chromosomal microarray analysis (CMA), is now being used to complement or as a replacement for traditional karyotyping. CMA is able to evaluate for smaller chromosomal deletions and duplications that can be seen by typical chromosome karyotype and can evaluate for microdeletion and microduplication syndromes that normal karyotyping cannot detect.

Microarray analysis can also detect additional genetic changes, including regions of homozygosity (ROH) or possible uniparental isodisomy. ROHs are long stretches of DNA that have similar polymorphisms or genetic loci that are similar. This can suggest consanguinity in the parents of the fetus because individuals, who are related, have a larger number of similar polymorphisms in their DNA. In couples who are consanguineous, there is an increased risk for autosomal recessive disorders. Therefore, additional screening of the parents or offspring may be warranted. Similarly, large areas of homozygosity can be associated with uniparental isodisomy (inheriting two copies of the same chromosome from one parent). In addition to concerns regarding autosomal recessive disorders, the finding of uniparental inheritance of chromosomal material can be associated with imprinting disorders.

An imprinting disorder occurs when an area of the genome that is coded to require a DNA contribution from each parent lacks biparental inheritance. Mechanisms that lead to an imprinting disorder include a deletion of that region of DNA, imprinting errors of the region, or uniparental inheritance of that region of DNA. There is an imprinted chromosome region on the long arm of chromosome 15, which is associated with Prader-Willi and Angelman syndromes. Prader-Willi syndrome is associated with hypotonia, developmental delay, and failure to thrive in infancy followed by excessive eating and morbid obesity as a child ages. Prader-Willi syndrome is caused by a lack of a paternal contribution of the critical area on chromosome 15 (15q11.2-q13). Conversely, Angelman syndrome is caused by a lack of maternal contribution of this critical area. Angelman syndrome is associated with severe intellectual disability and developmental delay, speech difficulties, gait ataxia, and a typical demeanor of overly exaggerated happiness that includes frequent laughing, smiling, and hand flapping (this was previously called the "happy puppet" syndrome due to this happy demeanor).

A limitation to microarray analysis is that it cannot detect structural anomalies that do not result in copy number changes, such as balanced chromosome translocations or rearrangements. If there is concern regarding a possible translocation, as in couples with multiple miscarriages, traditional karyotyping would be recommended. Also, if a microarray is consistent with a deletion or duplication of material that could be the product of a translocation, follow-up karyotype analysis should be considered.

Screening and testing does not always involve direct testing of the fetus. Carrier screening for common recessive and X-linked inherited disorders is routinely offered to women who are pregnant or who are contemplating pregnancy. Options for screening are varied and can be customized depending on the patient's individual family and personal histories. The American College of Obstetrics and Gynecology recommends offering carrier screening for cystic fibrosis (CF) and spinal muscular atrophy to women of reproductive age. Historically, carrier screening options have depended heavily on the patient's ethnicity. Hemoglobinopathy screening is recommended for individuals of African American, Asian, and Mediterranean background, as hemoglobinopathies are more common in individuals of these ethnicities. Hemoglobinopathies encompass genetic disease of the hemoglobin and include thalassemias and structural hemoglobin variants (such as sickle cell disease). Tay-Sachs disease has routinely been offered to individuals of Eastern European Jewish and French-Canadian background. Tay-Sachs disease is a neurodegenerative disorder in which the deficiency of the enzyme hexosaminidase A results in accumulation of certain lipids in the brain causing eventual spasticity and death in childhood.

As carrier screening has become less expensive, large pan-ethnic expanded screening panels have become available to screen for a range of disorders that are common among patients of many different ethnicities. The choice of the specific carrier screening panel to be used may be determined by the patient's clinical history. For example, if a pregnant patient is carrying a fetus with an increased nuchal translucency (determined by ultrasound measurement of fluid at the back of the fetal neck), a larger parental carrier screening panel may be warranted, as this ultrasound finding can be associated with several genetic syndromes including spinal muscular atrophy, Smith-Lemli-Opitz syndrome, and metabolic disorders. Screening the patient for these disorders may aid in the determination of a potential fetal diagnosis. In patients with a family history, carrier screening panels that include the disorder of concern may be considered to aid in risk reduction. For instance, if a patient has a family history of congenital adrenal hyperplasia or 21-hydroxylase deficiency, a carrier screening panel that includes the CYP21A2 gene could be performed to aid in carrier risk reduction or confirm carrier status. It should be noted that, as carrier screening is not diagnostic testing, the detection rates for variants are not 100%. In addition, some variants may not be reported. Therefore, carrier screening should not be performed in lieu of clinical diagnostic testing or to confirm a clinical diagnosis.

Results of genetic testing can modify care in the prenatal or immediate postnatal setting. In pregnancies diagnosed with Down syndrome, fetal echocardiogram and increased monitoring in the third trimester may be recommended, since babies with Down syndrome have an increased risk for heart defects and fetal loss. Newborns who have congenital adrenal hyperplasia are at increased risk for possible life-threatening salt-wasting crises. Prenatal diagnosis of these disorders can prompt additional evaluations to occur and may allow for early treatments to be initiated in the affected child.

Prenatal testing and screening can be of great benefit to families by allowing preparation for the care and treatment of their child with a genetic disorder. Prenatal testing gives parents and caregivers the opportunity to learn about their child's condition, reach out to parent support groups, make arrangements for additional care and therapy, and make family and friends aware of the condition. This preparation can make the post-birth time less stressful for both the parents and their families.

Pediatric genetics

In the pediatric setting, a larger number of patients who are suspected of having a genetic diagnosis for their medical concerns are undergoing whole-genome sequencing (WGS) or whole-exome sequencing (WES). This is especially true in patients who have had previous microarray and targeted gene panel testing with negative results. WGS refers to sequencing of an individual's entire genome. Because many parts of the genome are noncoding, and most disease-causing variants are in the exons, WES has also been employed by many providers. WES refers to sequencing the protein coding parts of the genome, or the exons.

These technologies use an individual's personal clinical history for evaluation and result interpretation. Results from WES or WGS typically return thousands of sequence variants. The results are then tailored to provide variant information based on a patient's specific phenotype. The actual variants that are reported are typically limited to genes that are associated with the clinical phenotype provided by the referring provider. Therefore, the clinical information given to the laboratory is very important and should be as complete and specific as possible to obtain the most accurate results.

While these technologies can provide a large amount of information, the meaning of this information may not be known. Some variants are able to be classified as either disease-causing or benign variants, which do not cause disease. However, many variants have not been reported previously or have conflicting information regarding their pathogenicity. The American College of Medical Genetics has produced recommendations for reporting information obtained through WES and WGS. These guidelines were developed to help laboratories and practitioners develop standards for interpretation of variants. These recommendations define a standardized terminology to describe the likely effect of the variant as pathogenic, likely pathogenic, uncertain significance, likely benign, and benign. Pathogenic and likely pathogenic variants have considerable evidence and data that suggest an abnormal outcome. Likely benign and benign variants have evidence that suggest they are non—disease-causing changes.

Variants of unknown significance (VUSs) are changes in the genome that have an unknown impact on gene function. These include variants that have not been reported before or do not have enough clinical data to determine if the variant is associated with disease. Since the clinical outcome for a VUS is not known, it is typically not recommended to make treatment decisions based on the presence of a VUS.

When a VUS is found, additional testing may be recommended to help determine its significance. For instance, if a VUS is found in a child with polydactyly, microcephaly, and intellectual disability, then testing of additional family members may be recommended to help with variant evaluation. This testing may be even more useful if it is performed on similarly affected family members. The results of this additional testing may suggest that this variant is in fact pathogenic if it is seen in similarly affected members of the family and is not seen in unaffected relatives. The variant could be considered more likely to be benign if multiple unaffected family members have the same variant as the proband. Variant classifications can change over time based upon data collected from these family studies and additional information gathered about a variant's clinical presentation.

As the understanding of genetic diseases increases, treatments for many disorders are becoming more readily available depending on the specific variant involved. Duchenne/Becker muscular dystrophy is an X-linked disorder caused by variants in the gene that codes for dystrophin. Individuals with absent dystrophin have Duchenne muscular dystrophy (DMD). Affected boys with DMD usually show signs of muscle weakness by age 3—5 years, are in need of wheelchair assistance by age 12 years, and die in their 20s. In individuals who have Becker muscular dystrophy (BMD), dystrophin production is reduced or abnormal. Boys with BMD usually have a later age of onset and much longer life span. Most individuals with DMD have a deletion of exons in the dystrophin gene, which alters the reading frame of the gene. The reading frame is the exact sequence of amino acid triplets needed to have correct translation of a gene and production of the gene product. It is typically thought that variants that do not alter the reading frame of the DMD gene are associated with the less severe BMD, whereas pathogenic variants that affect the reading frame of the DMD gene cause dysfunction of the production of dystrophin.

A current therapy for DMD involves a technique called exon skipping during premessenger RNA production. This technique is designed to correct the reading frame and induce the ability to produce some dystrophin. This treatment is only effective for variants that include certain exons. Depending on the exact variant present, this therapy may or may not be available for patients. Therefore, knowing a patient's exact variant is helpful in assessing what treatments and therapies are available for them.

Adult and specialty clinics

In the adult specialty clinics, precision medicine has been used for many years in the cancer setting. Tumors from individuals with Lynch syndrome (an inherited colon cancer syndrome) tend to have microsatellite instability (MSI), which is characterized by multiple anomalies in areas of repetitive DNA sequences called microsatellites. Individuals who have tumors that show MSI-high (MSI-H), a high level of instability in the microsatellites in genes associated with Lynch syndrome, have an increased risk for germline pathogenic variants in these genes. Therefore, germline genetic testing on these individuals is typically recommended. Immunohistochemistry (IHC) is a complementary testing strategy to MSI testing. Most MSI-H tumors show a loss of protein expression for at least one of the genes associated with Lynch syndrome. Loss of expression of proteins within the tumor is helpful in identifying which corresponding genes to target for mutation analysis. Although with the reduced cost and increasing availability of multiple gene panels, targeting one gene for testing is not as common as in the past. Current guidelines by the National Comprehensive Cancer Network recommends MSI and/or IHC testing on all colon and endometrial tumors be performed as routine screening. Individuals with positive results are then offered further genetic testing.

Genetic testing of tumors, called somatic testing, can be used to help determine prognosis, diagnosis, and treatment strategies, which are more likely to be beneficial to the patient. Cancer is caused by many types of variants across a large number of genes. Cancer cells are genetically unstable and, therefore, acquire many variants as the tumor and cancer progresses. Some variants may provide an advantage for the cancerous cells to survive, whereas others have no selective advantage. Knowledge of the genetic changes that are present in somatic tumors can helpful in determining if a specific chemotherapy treatment is likely to be effective. Genetic testing may also be able to help clarify the type of tumor involved or if a cancer is more likely to recur in the future.

Traditional chemotherapy broadly affects both healthy and cancerous cells. Targeted therapy is designed to affect only certain genetic changes (targets), which allow cancers to survive and grow. Genetic testing helps to determine if certain variants within the tumor are associated with a better response to specific chemotherapy regimens. As a result, the most effective and beneficial treatments for patients can be determined. This can be especially true for patients who have had recurrence of disease or have not responded well to first-line therapies.

There are many types of cancer with known variants that have FDA-approved treatments. Detection of these variants allows for targeted treatment of the patient with approved drugs. For instance, Afatinib targets EGFR (specifically exon 19 deletions or exon 21 substitution variants) in non–small-cell lung cancer. Other variants are known to lead to resistance to certain treatments, which then should be avoided.

Somatic genetic profiles may also show prognostic information, which may affect treatment decisions. In a patient with acute myeloid leukemia with poor prognostic genetic markers, a decision to progress to bone marrow transplant instead of chemotherapy may be considered by the treating physician. Somatic genetic testing may also help distinguish if a patient is a candidate for alternate therapies or research trials.

Germline genetic testing for other genetic disorders can also affect treatment options. Cystic fibrosis (CF) is one of the more common genetic disorders that has mutation specific treatments. CF is an autosomal recessive multisystem disorder that affects the epithelia of the respiratory tract, exocrine pancreas, intestine, hepatobiliary system, and exocrine sweat glands. It is most commonly associated with obstructive lung disease and malnutrition from pancreatic insufficiency. CF is caused by pathogenic variants in the CFTR gene. The normal CFTR protein is made in the endoplasmic reticulum and then moves to the surface of the cell to become a functioning salt channel. CFTR variants are broken up into six classes dependent upon the specific functional anomaly of the protein (Fig. 5.1). Delta F508 is the most common pathogenic variant associated with CF. It is a class 2 variant that results in incorrect protein folding. As a result, the protein cannot leave the endoplasmic reticulum and is unable to reach the surface of the cell. Class 3 mutations lead to proteins that can travel to the surface of the cell but cannot make a functioning salt channel. G551D is the most common class 3 gating variant. These two classes of variants have specific treatments, which are designed to target the processing of the anomalous protein.

Treatments have been developed, which help to open the salt channel once the protein reaches the surface of the cell and aids with ion transport. For individuals who have homozygous delta F508 variants, therapies have been designed to help

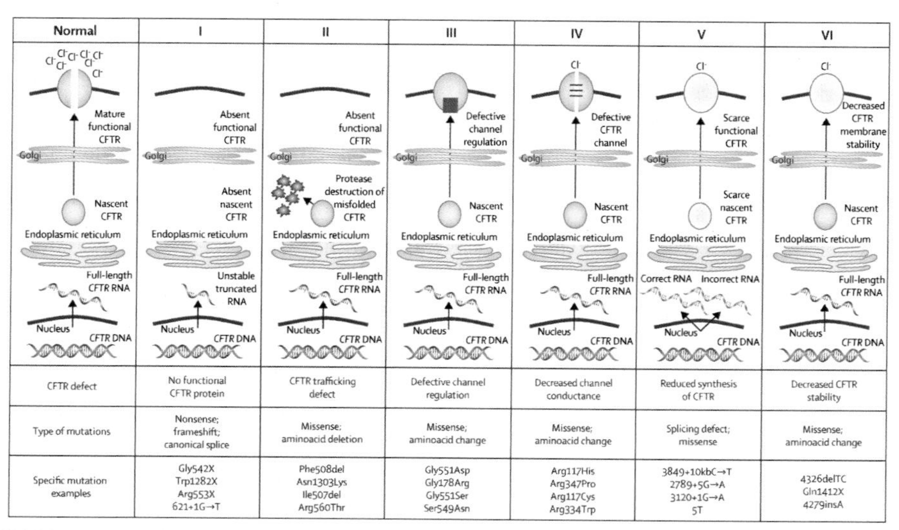

FIGURE 5.1

Cystic fibrosis variant classifications.

fold and move the protein to the surface of the affected cells. Some patients may benefit from treatment with multiple interventions. The use of precision medicine has made a significant impact on the treatment plan for patients who have CF.

Cardiologists also use genetic testing to help guide treatment and surveillance. There are several types of autosomal dominantly inherited cardiac syndromes including familial cardiomyopathies and cardiac rhythm disorders. Diagnosing a pathogenic variant in an affected individual can confirm a genetic etiology and allow for clinically unaffected family members to have genetic testing performed. This testing can guide surveillance recommendations for the entire family. A majority of hypertrophic cardiomyopathy (HCM) is familial and caused by an underlying genetic syndrome. After a proband has been diagnosed with clinical HCM, longitudinal cardiac surveillance is recommended in clinically unaffected first-degree family members. However, if a known pathogenic gene variant is found in that proband, which confirms a genetic etiology for the HCM, follow-up genetic testing of relatives can be initiated and can guide medical management for these individuals.

There are obvious advantages to tailoring treatments to a patient's specific genotype when dealing with rare genetic disorders or cancers. On the other hand, the use of precision medicine on the healthy population brings up many additional questions. Currently, there is debate about the utility of offering WGS to the general population.

WGS can evaluate for the presence of variants in multiple genes, which are associated with disease. The results of this testing may help to determine if surveillance for a previously undiagnosed disorder should be initiated. For instance, if a patient is found to have a variant in a gene associated with cardiomyopathy, cardiac evaluation would be recommended.

Polygenic or multifactorial disorders such as diabetes, heart disease, and obesity have a significant genetic component to their etiologies. However, it has historically been difficult to determine a specific genetic variant that can help stratify a patient's risk to develop these common disorders. Therefore, there is interest in using WGS or WES to evaluate multiple genetic alterations associated with risks for polygenic disorders. Risk stratification could then be based on the presence or absence of these alterations.

One difficulty in using this protocol is that risk stratifications are not designed to evaluate for all possible traits that can affect the likelihood for a person to develop these polygenic diseases. Hence, a low-risk evaluation does not eliminate the chance for a person to develop the disease in question. In addition, individuals of rare ethnicities may have a less accurate risk assessment, since many early research studies had a lack of diversity in their genetic samples. Patients must also be aware that many variants can have a range of clinical outcomes (known as variable expressivity). In addition, for many disorders not all individuals who carry a pathogenic variant will have symptoms (known as reduced penetrance). As a result, even if a pathogenic variant is present, the risk for development of the disease is frequently not 100%.

The use of WGS in the general healthy population can give an abundance of new information regarding genetic risk factors. This information can be important for patients who do not have the luxury of a detailed family history available to help with risk assessment. Patients who are adopted or are estranged from their families may not have family history knowledge for clinicians to use and evaluate. In these and many other cases, the information that WGS or WES can provide could be advantageous in guiding primary care.

One guiding principle to consider when debating the use of WGS in the general population is that testing should lead to a medically actionable plan. Will results change care or improve medical care for the patient? In some cases, the answer is yes. If a provider finds that a patient has an increased risk for colon cancer, then the provider can increase cancer surveillance, recommend lifestyle changes, and make the patient aware of early clinical symptoms to allow for early detection. In other cases, the answer is no. This may be especially true for disorders that do not have any available treatments or risk-reducing alternatives available.

Alpha-1-antitrypsin (AAT) deficiency is an autosomal recessive condition that is associated with an increased risk for emphysema and liver disease. AAT is a major inhibitor of neutrophil elastase in the lower respiratory tract. Decreased levels of AAT leave individuals vulnerable to progressive destruction of alveoli by neutrophil elastase, which leads to emphysema. In addition, AAT deficiency causes liver disease in some patients. Individuals with pathogenic mutations in the AAT gene have a substantial risk for obstructive lung disease (particularly if they are smokers). Therefore, these individuals should be counseled to avoid smoking and employment in professions that would lead to smoke inhalation or contact with asbestos. Awareness of these variants may make a very significant change in the patient's well-being.

There is also concern about potential harm that could come from offering WGS to healthy individuals. After receiving their results, patients may then undergo additional diagnostic tests or treatments that are unnecessary. Hemochromatosis is a very common recessive disorder that can cause iron overload in affected individuals. Approximately 1 in 10 Caucasians are carriers of hemochromatosis, and the vast majority of affected individuals will not have clinical symptoms in their lifetime. If a nonsymptomatic patient is found to have two variants for hemochromatosis, is it prudent to initiate screening for iron overload? Should the patient be advised to have prophylactic chelation therapy? Typically, treatment is initiated based upon symptoms. Is it appropriate to defer the initiation of treatment in a patient with a known genetic disorder who does not yet have symptoms? There may not be a right or wrong answer.

Biotinidase deficiency is a recessive metabolic disorder that causes the body to be unable to reuse the B vitamin named biotin. This condition is treatable in affected infants and children by giving biotin supplementation. However, if untreated, biotinidase deficiency can cause delayed development, seizures, hypotonia, breathing problems, hearing and vision loss, and other difficulties. The c.1330G>C variant is a common mild variant that does not result in a disease phenotype even when homozygous. Therefore, individuals who are found to carry two copies of this particular variant are not at risk for disease. A homozygous result for the c.1330G>C variant may prompt additional testing and screening to evaluate for clinical biotinidase deficiency. This additional testing would be considered unnecessary.

The issue of performing potentially unnecessary additional screening and testing is especially true when faced with variants of unknown significance. The presence of a VUS may prompt physicians to order additional tests or screening measures in clinically normal individuals. However, as these genetic variants have unknown clinical utility, their presence should not dictate medical management. If unnecessary follow-up is ordered, it leads to concern about avoidable medical spending and wasteful use of resources.

In addition, WGS and WES can detect many disorders that are not expected. Dealing with these unexpected results is a major apprehension when considering the benefit of offering these test options. However, it does not outright suggest that WGS should not be performed in the healthy population. It means that appropriate counseling and follow-up should be in place to answer potential questions from patients regarding test results.

While genetic testing is not always perfect and the results may be difficult to interpret, these results can be used to help guide medical management. The best care plan for the patient should be developed using genetic test results in conjunction with all other available clinical information.

Genetic counseling

One of the first definitions of genetic counseling was made by the American Society of Human Genetics in 1975 and was interpreted as "a communicative process that deals with the human problems associated with the occurrence, or the risk of occurrence, of a genetic disorder in the family." The scope of genetic counselors has exponentially increased in the past 40–50 years. This underlying definition is still a major part of the genetic counseling profession, but genetic counseling has grown to encompass a much broader range of job roles. Currently, genetic counselors work in varied settings including medical centers, universities, laboratories, government agencies, and genetic disease support organizations.

Genetic counselors are healthcare professionals with specialized graduate degrees and experience in medical genetics and counseling. Genetic counselors are trained in human genetics, embryology, lab science, ethics, counseling techniques, genetic testing methods, and results interpretation. They are uniquely qualified to support and integrate precision medicine into their practices, both in the pretest and in the posttest decision-making process. In the clinic, genetic counselors can serve many roles including reproductive and familial risk evaluation, facilitation of pretest decision-making, coordination of testing, results interpretation, and management of patient perceptions and expectations.

Genetic counselors are trained to look closely at family and personal histories to determine the options for genetic testing, which may best benefit the patient. For instance, in a patient with a personal history of ovarian cancer, testing for hereditary breast and ovarian cancer syndromes may seem to be the best pathway of testing. However, if there is a significant family history of uterine or colon cancers, the focus may shift to a concern for Lynch syndrome. As a result of asking more targeted questions regarding family history, additional testing for hereditary colon cancer genes may also need to be ordered.

One common component of the genetic counseling process is the construction of a family pedigree, which is a pictorial representation of the patient's family medical history. It looks figuratively at this information and describes affected individuals, age of onset of disease, number of siblings, and number of generations affected, among other details. Pedigrees allow the provider to look at a large amount of information in one document and to evaluate that data in a more precise manner. These pedigrees can make determining an inheritance pattern or possible working diagnosis easier to assess and visualize. The pedigree can also be used to show the patient which medical information is of concern within the family and to explain why an inherited or genetic disorder is suspected.

In the attached pedigree (Fig. 5.2), patient B1 is referred for prenatal genetic counseling due to a history of cleft palate in her son (C1). Isolated cleft palate has a multifactorial etiology in many cases. Evaluation of the rest of the family history reveals that there is also a family history of congenital heart disease in the patient's father (A1) and nephew (C6), and the patient's brother has intellectual disabilities (B3). This additional information allows the genetic counselor to consider alternative genetic syndromes, which may explain this history, such as the 22q11.2 deletion syndrome. This microdeletion syndrome, also known as

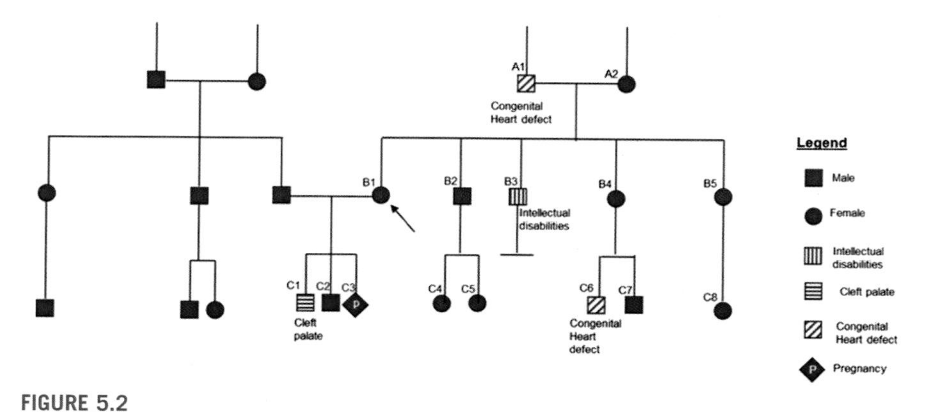

FIGURE 5.2

Pedigree for a patient referred for a history of cleft palate in her son. Notice that there is additional history of congenital heart disease and intellectual disabilities which could suggest a possible familial syndrome such as 22q11.2 deletion syndrome.

DiGeorge syndrome, is an autosomal dominant disorder, which has variable presentation. It is associated with several anomalies including congenital heart defects, cleft palate, intellectual disabilities, kidney anomalies, and immune disorder. This constellation of findings within a family would prompt many genetic counselors to recommend evaluation for 22q11.2 deletion syndrome.

Genetic counselors are experts who are skilled at discussing both the social and medical concerns that patients may have regarding genetic testing. There are laws in place that protect individuals in the United States from genetic discrimination. For instance, the Genetic Information and Nondiscrimination Act of 2008 (GINA) is a federal law that protects individuals against the use of their personal genetic information to discriminate against them in health insurance and employment areas. GINA does have limitations, however. It does not apply to individuals in the military and does not address the possibility of genetic discrimination by life insurance companies. Therefore, patients may have concerns regarding consenting to genetic testing due to possible discrimination in the future. Genetic counselors can address these concerns and possible options for the patient.

Patients may also feel anxious about learning the results of their own testing: how will they cope if the result is positive—what if the result is negative? Parents may experience feelings of guilt if they are at risk to pass a genetic variant to their children. Genetic counselors play an integral role in discussing these issues and the pros and cons of performing testing.

The thought of genetic testing can be overwhelming for patients, especially if they have little medical knowledge or understanding of genetic concepts. Genetic counselors take very complex genetic and medical concepts and explain them at a level that patients can better understand. As an example, the genome can be explained as a human genetic library with each chromosome being complementary to a row of books. Each gene is then analogized as a book, which needs to be accurately read for the correct proteins to be made. Variants are described as changes in the text of the book, which may affect how the body "reads the book" and therefore makes proteins. Genetic counselors are able to explain difficult concepts using a variety of strategies to help patients understand the basic concepts of inheritance, genetic disease, and testing.

Genetic counselors give patients the information and support that they need to make informed decisions regarding testing. As these conversations may be very time-consuming, this can become a challenging task for physicians in an era where the typical time allotted for a physician consult can be as little as 15 min. A genetic counselor is a skilled medical professional who is an invaluable member of the patient care team and is an excellent choice to provide pretest counseling to a patient who is a candidate for genetic testing.

In addition, genetic counselors can provide significant insight into the choice of genetic testing laboratories. Genetic counselors are often the healthcare providers who meet with the laboratory representatives and have the latest information on changes to laboratory testing protocols and test menus. Genetic counselors are experts at evaluating the available testing options to determine the best option based upon the patient history, needs, and medical insurance. They also work as liaisons between the laboratory and clinic to help with specimen procurement, coordination, and provision of clinical history to the laboratory. This special skill set makes the testing process more seamless for the patient and the provider.

Posttest genetic counseling is also quite important, because it allows the provider to explain the test results in detail to patients and their families, including discussion of the ramifications of testing. This dialogue may include explanation of risk values, penetrance information, and recommended surveillance based on testing results. There are many genetic variations that do not impose 100% risk for development of the disorder on the affected individual. The type of variation, age of the patient, and penetrance of the disorder all have an impact on the potential for disease. Patients should be made aware of these risks and limitations to result interpretation and should have a discussion regarding the options for treatment and/or surveillance.

Genetic testing is unique in that evaluation and results can have an impact on the entire family. Most medical testing is geared toward diagnosis and treatment of just the patient. Genetic testing results may give specific and useful information that can be used by the whole family. Therefore, dissemination of these results to the family is of utmost importance, with many genetic counselors offering the option of a "family results" letter that can be shared with family members of the patient. These letters provide information that allows family members to seek their own medical evaluations, without putting the burden of explaining medically complicated information on the patient. Genetic counselors can expedite the process of evaluation and testing of additional family members by acting as a central coordinator for familial testing.

Coordination of familial testing is paramount in families with familial hypertrophic cardiomyopathy. In these families, one affected individual is typically tested as the proband, with additional family members having cascade testing performed if a pathogenic variant is discovered. Cascade testing involves testing of additional family members once a pathogenic variant is found in the proband. Genetic counselors can help to determine which family member is the best candidate for initial testing, depending on clinical symptoms, age of onset, or even insurance limitations.

Many patients expect genetic testing to give them all of the information that they desire to learn regarding their risk for disease and options for treatment. An integral part of the genetic testing process includes discussion of the patient's perception and expectations of what they will gain from testing. Genetic testing results may not always provide the answers desired by the patient or providers and, in some cases, may lead to additional uncertainty for the patient. Managing patient's expectations and educating patients about possible outcomes of testing will increase patient satisfaction and reduce disappointment and frustration with testing.

Further reading

1. ASHG – American Society of Human Genetics Ad Hoc Committee on Genetic Counseling. Genetic counseling. *Am J Hum Genet* 1975;**27**:240–2.

2. Nadler HL. Antenatal detection of hereditary disorders. *Pediatrics* 1968;**42**(6):912–8.
3. Richards S, et al. Standards and guidelines for the interpretation of sequence variants: a joint consensus recommendation of the American College of medical genetics and genomics and the association for the molecular pathology. *Genet Med* 2015;**17**(5):405–24.
4. The American College of Obstetricians and Gynecologists. *Carrier screening for genetic conditions: committee opinion number 691*. 2017.
5. Vanderbilt-Ingram Cancer Center. *My cancer genome* [online]; 2010–2019 https://www.mycancer_genome.org.
6. Elborn JS. Cystic fibrosis. *Lancet* 2016;**388**(1059):2519–31.
7. Univeristy of Washington. *Gene Reviews: Biotinidase Deficiency* [online] Seattle (WA): University of Washington, Seattle; Updated June 6, 2016 https://www.ncbi.nlm.nih.gov/books/NBK1322.
8. Univeristy of Washington. *Gene Reviews: Dystrophinopathies* [online] Seattle (WA): University of Washington, Seattle; Updated April 26, 2018 https://www.ncbi.nlm.nih.gov/books/NBK1119.
9. Univeristy of Washington. *Gene Reviews: Alpha-1-Antitrypsin deficiency* [online] Seattle (WA): University of Washington, Seattle; Updated January 19, 2017 https://www.ncbi.nlm.nih.gov/books/NBK1519.
10. National Comprehensive Cancer Network. *NCCN Guidelines Version 1.2018 Lynch Syndrome* [online]; Updated 7/12/2018 https://www.nccn.org/professionals/physician_gls/pdf/genetics_colon.pdf.

CHAPTER

Pharmacogenomics

6

Judy S. Crabtree, PhD [1,2]

Associate Professor, Department of Genetics, Louisiana State University Health Science Center, New Orleans, LA, United States[1]*; Scientific and Education Director, Precision Medicine Program, Director, School of Medicine Genomics Core, Louisiana State University Health Science Center, New Orleans, LA, United States*[2]

Chapter outline

Introduction

With the completion of the human genome, we are learning more and more about how variation in the genome can influence modern medicine. This increased understanding has led to a plethora of new fields, the so-called "-omics technologies," and includes pharmacogenomics, which combines the field of pharmacology (or the science of drug uses, effects, and modes of action) with genomics (the study of genes, gene variants, and gene function). Pharmacogenomics seeks to understand how genes and gene variation influence an individual's response to pharmaceutical drugs. Currently, most drugs on the market are "one size fits all" and do not take into account the genetics of the patient, or in the case of oncology, the unique genetics of the patient's tumor. It is challenging to predict who will respond well and who will not, and who will experience side effects and who will not. The goal of pharmacogenomics is to understand the interaction between drugs and genomics such that more specific drugs can be developed and the right drug(s) can be given to the correct patient population (i.e., the population with a particular genetic signature) to

Clinical Precision Medicine. https://doi.org/10.1016/B978-0-12-819834-6.00006-9

maximize efficacy while minimizing side effects. This concept is the foundation of precision medicine—the right drug, given to the right patient, at the right dose, at the right time.

The field of pharmacogenomics is still in its infancy with much more to learn. Some major breakthroughs in pharmacogenomics are incorporated into regular clinical care, but these cases are few. Many pharmacogenomics approaches are still in development or in clinical trials, with a focus of moving these approaches toward translation to active clinical decision-making. Furthermore, for these discoveries to be translated to the clinic, the structure of clinical trials must change, and the genomics infrastructure for clinical testing must also grow. Clinical genomics laboratories are becoming more commonplace in major medical centers across the United States and the world, as variants associated with both disease and drug responses are cataloged, tested, and moved into clinical care. See Chapter 8 for a discussion of the technologies behind genetic testing.

DNA variants

There are two general types of DNA variants. Germline variants are changes in the DNA sequence that are inherited from the mother or father and can be responsible for natural variation or disease pathology. Somatic variants are changes that were not inherited but which occurred only in the individual. Furthermore, DNA variants can be large chromosomal aberrations (translocations, chromosomal loss, etc. as described in Chapter 1), large insertions/deletions, or single nucleotide polymorphisms (SNPs). As we are learning from the 1000 Genome Project, SNPs are very common in the human genome, on average one every 300 nucleotides, or roughly 10 million SNPs in the whole human genome.

SNPs are often responsible for changes in function of the resulting gene product, be it a functional RNA or a protein. Long noncoding RNAs with variants may disrupt the normal complementarity with targets, which alters the tethering function of lncRNAs, potentially recruiting accessory proteins to incorrect sites within the DNA. In the case of proteins, SNPs can be either a structural variant or an expression variant, depending on the location of the SNP within the gene. Structural variants result in changes in the three-dimensional structure of a protein. These changes may be minor, or they may significantly change substrate specificity and alter binding partner interactions if the variant is present in a critical domain of the protein. Expression variants change the amount of protein/enzyme that is produced in the cell and are usually the result of variants in the promoter region or splice sites of genes. Expression variants often change the ratio of enzyme to substrate or alter concentration of the protein with respect to its binding partners within the cell.

In addition to the genomic databases that catalog the presence of SNPs in the human genome, there are also databases that catalog the correlation of SNPs with human disease (e.g., ClinVar, https://www.ncbi.nlm.nih.gov/clinvar/) and other databases that correlate SNPs with drug efficacy and adverse drug events (e.g.,

PharmGKB, https://www.pharmgkb.org/). As these databases continue to grow, our ability to correlate and predict drug efficacy with respect to presence or absence of SNPs (known as a SNP map or profile) will significantly improve. These data also inform clinical trials, allowing for more appropriate clinical trial design based on genetics to ensure the appropriate patients are included in the trial for maximum efficacy and fewer adverse events. There are also national programs in the United States and worldwide to learn more about SNP variation between individuals, within ethnic groups, and across geographical regions (e.g., All of Us; https://allofus.nih.gov/).

What does identification of DNA variants mean to patient care? Ultimately, if pathogenic variants can be identified *before* a patient develops a disease, controllable factors such as diet, exercise, smoking, weight gain/loss, and environmental toxin exposures can be modulated to minimize risk of developing the disease. If variants are identified before a patient receives a drug, the best drug can be selected for the patient (eliminating the trial and error approach to prescribing), minimizing side effects and improving clinical outcomes.

DNA variants and pharmacology

DNA variants can influence drug pharmacodynamics and pharmacokinetics in many ways. Within the training for clinical personnel is the inevitable study of drug pharmacology that includes absorption, distribution, metabolism, and excretion (the ADME principles) of any drug that enters the body. Prior to FDA approval, drug development companies must demonstrate how their drug is activated (if a prodrug) or metabolized, how it gets absorbed into the relevant tissue, where else the drug and/or its metabolites accumulate in the body, and how the body removes the drug and/or its metabolites. Genetic variation in any of the ADME parameters is often the cause of treatment failure and patient nonadherence to a drug schedule due to a negative side effect profile. Interestingly, differences in side effect profiles that correlated with ethnicity were the first indication that a particular drug response could be inherited and that inherited variants in DNA sequence may be responsible for some clinical outcomes.

Metabolic processes in the liver are the primary mechanisms of drug metabolism, also called the "first pass" effect. The cytochrome P450 enzymes in the liver are responsible for 80% of phase 1 metabolism through reactions that break down drugs into active or inactive metabolites. There are approximately 60 cytochrome P450 genes in humans, and variants have been identified in the majority of these genes. The genes that encode these proteins have a particular nomenclature that identifies the family, subfamily, gene number, and if relevant, a variant, which is called a star allele. For example, the gene *CYP2D6*10* is the family CYP2, subfamily D, gene number 6, and star allele 10, which makes the CYP2D6*10 protein. The different families of CYP enzymes interact with different classes of drugs based on the structure of the drug. Structural variants can alter the active site of these enzymes, changing which drugs are accessible to the enzyme and impacting metabolic rate and/or specificity.

CYP enzymes can be induced or inhibited by other substances in the body such as other drugs or dietary supplements, which adds an additional layer of complexity to CYP enzyme regulation. For example, St. John's Wort, a commonly used dietary supplement thought to improve symptoms of mild depression, is an inducer of CYP3A4. In this scenario, overactive CYP3A4 can cause rapid metabolism of drugs cleared by CYP3A4, decreasing their effectiveness. Grapefruit juice works in the opposite way and inhibits CYP3A4, which may cause slow or absent metabolism of drugs, increasing exposure with potentially dangerous side effects. Just as combinations of prescription or over-the-counter drugs, and dietary inducers or inhibitors, variants in CYP genes also alter CYP enzymatic activity and change the pharmacokinetics and pharmacodynamics of drugs.

CYP variants in anticoagulant therapies

Patients at risk for or after having a stroke are treated with anticoagulants to thin the blood. There are many of these medications on the market, but warfarin and Clopidogrel are excellent examples of the impact of genetic variation on CYP metabolism. Without any genetic analysis, the dosing of anticoagulant is trial and error with the physician modulating the drug choice and dosage until acceptable anticoagulant activity is achieved.

Warfarin

Warfarin is prescribed to patients to prevent thromboembolic diseases in patients with deep vein thrombosis, atrial fibrillation, recurrent stroke, or heart valve prostheses. Warfarin is produced and dosed as a racemic mixture (equal amounts of two mirror image compounds, termed R-warfarin and S-warfarin). Both R- and S-warfarin are metabolized by a variety of CYP enzymes. R-warfarin is metabolized by CYPs 1A1, 1A2, and 3A4 into an inactive metabolite that is rapidly cleared from the body, whereas S-warfarin is metabolized by CYP2C9 into the inactive 7-OH-warfarin metabolite. The primary function of warfarin is to inhibit the reaction of oxidized vitamin K to reduced vitamin K that is catalyzed by the vitamin K epoxide reductase enzyme, encoded for by the *VKORC1* gene on chromosome 16. This reduces the amount of vitamin K that is available to serve as a cofactor for clotting proteins in the blood clotting cascade.

Warfarin dosing is challenging due to a narrow therapeutic window and a high degree of interpatient variability. Patients on warfarin are subjected to repeated, regular testing to ensure proper drug function and anticoagulant activity. Warfarin interpatient variability is partially due to variants present in the CYP2C9 gene that alter the normal function of this enzyme. To date, more than 30 polymorphic alleles (SNPs) have been identified in the CYP2C9 gene. The *1 allele is the wild-type allele, also called the consensus allele, and is the allele to which all others are compared. In CYP2C9, the most prevalent alleles are the *2 and *3 alleles, which decrease CYP2C9 activity for warfarin by 30% and 90%, respectively. *2 is considered an intermediate metabolizer, and *3, a poor metabolizer of warfarin and genetic testing reports, will typically include descriptions with that nomenclature. Remember that individuals have two copies of every gene, one inherited from their father and one from their mother, so any given individual can have any combination of variant alleles, which accounts for some interpatient variability: *1/*1, *1/*2, *1/*3, *2/*2, *2/*3, or *3/*3. The *2 and *3 alleles require a 19% and 33% reduction in warfarin dose per allele. A key point to note is that this decrease in warfarin metabolism is a result of *decreased CYP2C9 enzymatic activity due to variants* and applies to *any* drug metabolized by CYP2C9, not just warfarin.

As a side note, there are also major variants identified in the *VKORC1* gene. G3673A is a variant in the promoter of the *VKORC1* gene and is also often referred to as the -1639 G>A allele or rs993231. This SNP alters a *VKORC1* transcription factor binding site, leading to a 44% decrease in VKORC1 expression when the A allele is present compared with the G allele. This SNP, like many others, tracks with ethnicity as well. This variant is present in 90% of individuals of Asian descent, 40% of Caucasians, and 14% of individuals of African descent.[1] The presence of a single A allele requires a decrease in warfarin dose of 28% per copy.

The spectrum of interpatient variability that is noted in patient responses to warfarin is in part due to the presence of any combination of these variant alleles. For simplicity sake, not all possible variants are discussed here, and others exist such as variants in *CYP4F2*, which further expands the options for variability among patient responses. Finally, if genetic testing has been performed for variants in *CYP2C9, CYP4F2,* or *VKORC1*, there are tools online to assist with warfarin dosing adjustment based on the results of genetic testing (www.warfarindosing.org).

Clopidogrel

Clopidogrel is another anticoagulant drug that is actively prescribed for patients at risk for thromboembolic events. Clopidogrel is dosed as a prodrug, and once in the liver, it must be metabolized by CYP2C19 into the active metabolite, which then functions to irreversibly inhibit the platelet ADP receptor P2RY12. Like warfarin, several other CYP enzymes act upon clopidogrel, but for simplicity, CYP2C19 will be the focus here. As with warfarin, the CYP2C19*1 allele is the wild-type allele and is considered the normal metabolizer of clopidogrel. CYP2C19*2 and *3 alleles are poor metabolizers, whereas CYP2C19*17 is a hypermetabolizer. Combinations of these variants lead patients to be classified into one of several categories: ultrarapid metabolizer (UM), extensive normal metabolizer (EM), intermediate metabolizer (IM), or poor metabolizer (PM) as shown in Table 6.1. Because PM patients treated with clopidogrel at recommended doses exhibit higher cardiovascular adverse events compared with patients with normal CYP2C19 function, the FDA requires a boxed warning on clopidogrel, which recommends genetic

Table 6.1 Variant combinations resulting in classification of patients into one of four categories: UM, EM, IM, or PM.

Variant(s)	Metabolizer type	Alleles
*17 het or hom	UM—rapid/ultrarapid metabolizer	*1/*17, *17/*17
*17 and *2 OR *3	Unknown status (one gain, one loss of function)	*2/*17, *3/*17
No variant present	EM—extensive "normal" metabolizer	*1/*1
*2 or *3 het	IM—intermediate metabolizer	*1/*2, *1/*3
*2 or *3 hom	PM—poor metabolizer	*2/*2, *2/*3, *3/*3

testing for CYP2C19 prior to prescribing. A boxed warning is the strongest warning from the FDA, signifying that the drug has significant risk of serious or life-threatening adverse events. As with warfarin, a key point to note is that changes in metabolism of clopidogrel are a result of *altered CYP2C19 enzymatic activity due to variants* and apply to *any* drug metabolized by CYP2C19, not just clopidogrel.

DNA variants in drug transporters

Drug metabolism is also affected by variants in genes that encode drug transporters. One such example is the SLC47A1/MATE1 transporter, which is responsible for excreting metformin into the bile. Metformin is an older, yet commonly prescribed medication for type II diabetes, particularly in obese or overweight patients. Metformin decreases blood glucose levels, low-density lipoproteins, and triglycerides, as well as increasing peripheral insulin sensitivity to help control type II diabetes. In the liver, metformin works by activating AMPK, inhibiting gluconeogenesis, and increasing hepatic insulin sensitivity. The SLC47A1/MATE1 transporter pumps metformin into the bile for excretion via the kidneys and urine. Unlike warfarin or clopidogrel, metformin is not acted upon by the CYP enzymes and is secreted unchanged.

Variants have been identified in the *SLC47A1* transporter gene by studying patients who responded well to metformin versus patients with no response.[2] A SNP called rs2289669, which is a G to A transition, had the highest frequency of the minor A allele in diabetics who responded well to metformin. Patients homozygous for the A allele also had the largest decrease in A1C levels after taking metformin and had a decrease in A1C of 0.87% compared with patients homozygous for the G allele.

Ultimately, these data mean that the A allele decreases or inactivates the SLC47A1/MATE1 transporter such that less metformin is excreted to the bile and more is available to stay in the liver to decrease glucose levels and increase insulin sensitivity. The result is improved diabetic control. As with anticoagulants, diabetic patients on metformin exhibit a full spectrum of responses from no response to highly successful responses—reflecting that the rs2289669 G>A SNP is not the only identified SNP that impacts diabetic control via metformin. Many more SNPs exist that impact metformin excretion and transport to varying degrees, which begins to explain the variability of responses to metformin at the pharmacological level. In the future, one could envision a SNP genotyping panel for all diabetes-related SNPs that would aid clinicians in determining the best therapeutic agent and dose prior to the start of any treatment.

DNA variants in enzymatic pathways

Drug disposition can also be impacted by variants present in pathways that act on drugs but are not a part of the CYP superfamily of enzymes or a transporter regulating cellular localization. A classic example of this mechanism is degradation of the chemotherapeutic 5-fluorouracil (5FU), which is a widely used chemotherapeutic for solid tumors that is often used in combination with other chemotherapeutic agents (Fig. 6.1). As its name implies, 5FU is an analog of uracil that negatively impacts the pyrimidine biosynthesis pathway and DNA/RNA synthesis. 5FU is dosed as a prodrug, and more than 80% of 5FU is inactivated and degraded in the liver by the enzyme dihydropyrimidine dehydrogenase (DPD). The remaining 20% of 5FU is converted to the active compound 5-fluoro-2-deoxyuridine monophosphate (5-FdUMP) in a series of reactions. Active 5-FdUMP inhibits thymidylate synthase within the pyrimidine biosynthesis pathway as well as inhibiting DNA/RNA synthesis and blocking cell division to kill rapidly dividing cancer cells.

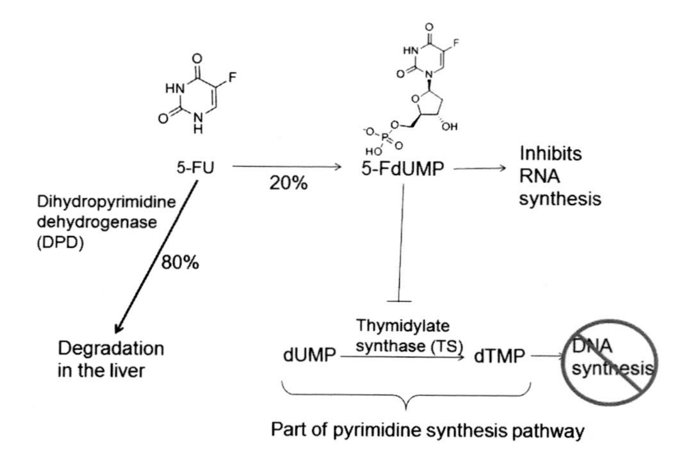

FIGURE 6.1

Mechanism of action of 5-fluorouracil.

Several SNP variants have been identified in *DPYD*, the gene that encodes the DPD protein. SNP variants that result in a nonfunctional allele such as DPYD*2A allele (rs3918290) and DPYD*13 allele (rs55886062) significantly influence chemotherapeutic toxicity. Patients heterozygous for one of these alleles are considered intermediate metabolizers of 5FU, whereas patients homozygous are poor metabolizers. Treatment with 5FU in patients with these variants has devastating consequences, including hematopoietic, neurological, and gastrointestinal toxicities that are often fatal. A recent review published in 2018 covers all the known genetic polymorphisms present in DPYD as well as the known toxicities associated with 5FU treatment in patients with these variants.[3] The incidence of individuals with DPYD variants is relatively low, but many of these patients die from toxicity that could have been avoided with a simple genetic test.

A variant has also been identified in the thymidylate synthase gene. This variant is not a SNP but a polymorphic variable tandem repeat in the promoter of the gene. Patients can have one, two, or three copies of the repeat and can be homozygous or heterozygous, which further explains the range of possible patient responses to 5FU. Patients who are homozygous for the triple repeat have higher thymidylate synthase activity, often making these individuals chemoresistant to inhibition by 5FU, resulting in poor response to therapy and disease progression. The combination of potential thymidylate synthase and DPD variants begins to explain the spectrum of tumor responses to 5FU.

Drug development and clinical trials

The majority of the drugs on the market today are "one size fits all" toward the general assumption that all individuals will respond the same way to the same drug and the same dose. As the examples above illustrate, we now know that this is not the case, and as more variants are identified that correlate with disease and drug outcomes, the process of drug development will need to adapt.

The identification of drug targets is moving away from the ages-old criteria of the "next blockbuster drug" that everyone should take and instead is moving toward a more tailored, precision approach by focusing on smaller populations with a given variant. Clinical trial design is also adapting to this new paradigm and incorporating the concept that the same variant may impact various different disease states. For example, in oncology, the same variant may be present in cancers from multiple different tissue types.

The changes in drug design and development, particularly in oncology, have led to new clinical trial designs called basket and umbrella trials that are based on genetic testing of SNPs or biomarkers. Basket trials are a master protocol wherein each subtrial enrolls multiple tumor types and tests one variant or one drug in many cancers. Umbrella trials are the reverse, enrolling only one tumor type, but testing many potential variants and/or targeted drugs. The key benefit of umbrella and basket trials is that it is possible to identify options for smaller subgroups of patients, even with a variant that is present only in 1%–2% of the broad cancer patient population. Furthermore, this provides a tractable plan of treatment for patients with these rarer variants, and many times, patients can participate in trials for their specific variant without having to travel to distant sites. All of these approaches facilitate testing, regulatory approval, and clinical translation of new therapeutics to the clinic.

Conclusions

Pharmacogenomics, the interface of genomic testing and pharmacology, has great potential for improving medical management of patients through improved risk assessment, drug safety profiles, and drug development. If initial studies are any indication, the broad spectra of patient responses to drug therapy often can be traced to genetic roots, with DNA variants aligning with efficacy, side effects, and patient survival. Furthermore, as more drugs are scrutinized with respect to genetic variation and outcomes, we may see a resurgence of older drugs previously deemed "unsafe" due to side effects in a given population, but which were remarkably successful in others. Future drug development tailored to particular SNP profiles, and clinical trials designed to capitalize on these SNP biomarkers hold significant promise for future targeted therapies.

References

1. Owen RP, et al. VKORC1 pharmacogenomics summary. *Pharmacogenetics Genom* 2010; **20**(10):642–4.
2. Becker ML, et al. Genetic variation in the multidrug and toxin extrusion 1 transporter protein influences the glucose-lowering effect of metformin in patients with diabetes: a preliminary study. *Diabetes* 2009;**58**(3):745–9.
3. Lunenburg C, et al. Prospective DPYD genotyping to reduce the risk of fluoropyrimidine-induced severe toxicity: ready for prime time. *Eur J Cancer* 2016;**54**:40–8.

CHAPTER

7

Immunogenomics: steps toward personalized medicines

Fokhrul Hossain, PhD [1], Samarpan Majumder, PhD [2], Lucio Miele, MD, PhD [3,4]

Postdoctoral Researcher, Louisiana Cancer Research Center (LCRC) and Department of Genetics, School of Medicine, Louisiana State University Health Sciences Center, New Orleans, LA, United States[1]; Instructor – Research, Louisiana Cancer Research Center (LCRC) and Department of Genetics, School of Medicine, Louisiana State University Health Sciences Center, New Orleans, LA, United States[2]; Director for Inter-Institutional Programs, Stanley S. Scott Cancer Center and Louisiana Cancer Research Center, Louisiana State University Health Sciences Center, New Orleans, LA, United States[3]; Professor and Department Head, LSU School of Medicine, Department of Genetics, Louisiana State University Health Sciences Center, New Orleans, LA, United States[4]

Chapter outline

Introduction

Next-generation sequencing (NGS)–based immunogenomics approaches can be applied to understand the pathogenesis of autoimmune diseases, immune responses following vaccines, organ transplantation rejection, graft-versus-host disease, food allergy, and cancer treatments (Fig. 7.1). This chapter focuses on the current status of immunogenomics applications in oncology. Cancer immunogenomics describes the interaction between cancer cells, stromal cells, and immune/inflammatory cells through the parallel study of these cell types in tumor-bearing patients and animal models. The process of carcinogenesis includes a dynamic interaction between

Clinical Precision Medicine. https://doi.org/10.1016/B978-0-12-819834-6.00007-0

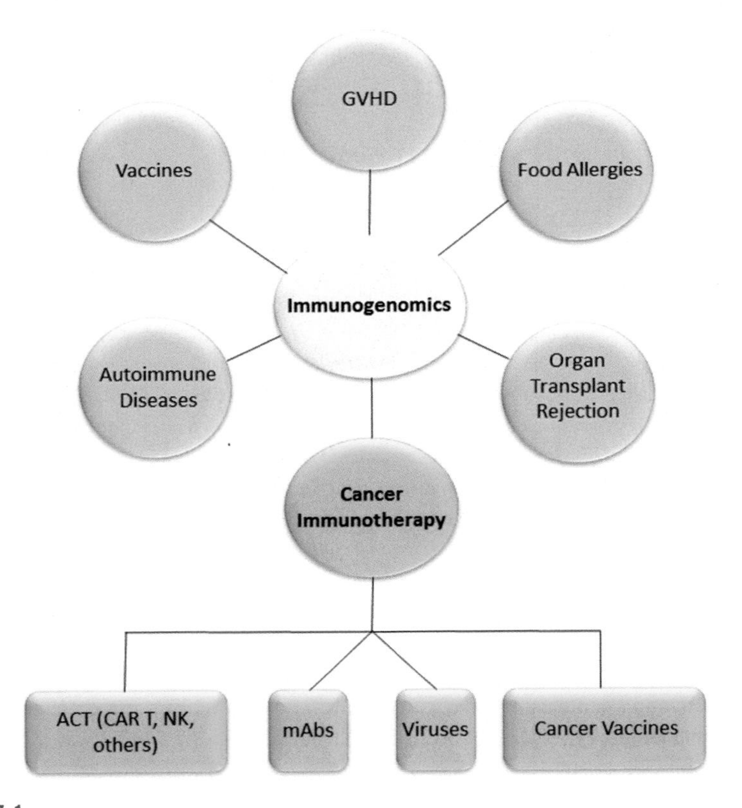

FIGURE 7.1

Applications of immunogenomics. The concept of immunogenomics can be applied to understand immune response–related disease processes including autoimmune disease (graft-versus-host disease [GVHD]), organ transplantation rejection, vaccines, food allergies as well as cancer immunity. This knowledge can be applied in the development of immunotherapeutics (monoclonal antibodies [mAbs], CAR-T cells and other adoptive cell therapies [ACTs], vaccines, and viruses).

the tumor and the immune system.[1,2] An immune surveillance phase whereby the immune system eliminates a majority of cancer cells is followed by an equilibrium phase during which cancers evolve immune-suppressive mechanisms and loses antigen-presenting major histocompatibility complex (MHC) receptors (immune editing). Eventually, a full-blown cancer escapes immune surveillance. Different cancers have different immunogenicity and different immune-suppressive activities. While all cancers contain some mutant genes, some cancers with DNA repair defects accumulate significant mutational burdens throughout the genome. Some mutations give rise to mutant proteins (neoantigens), which are potentially immunogenic. Mutated peptides/proteins are processed inside tumor cells and then presented at the cell surface bound to MHC class I receptors. This in turn triggers

T cell—mediated immunity. Mounting evidence suggests that tumor mutational burden (as measured by genomic NGS) correlates with the presence of neoantigens and predicts response to immunotherapy.[3–9]

Modern immunogenomics developed from a huge body of research led by pioneering cancer immunologists during the late 1980s through the 1990s. Once it was recognized that most cancers were of nonviral origin,[10–12] the identification of tumor-specific "neoantigens" on the tumor cell surface, bound to HLA molecules followed.[11,13] The identification and characterization of the role of MHC proteins in antigen presentation moved the cancer immunology field forward.[14,15] Successively, a method to culture antigen-specific cytolytic T lymphocytes (CTLs) was developed,[16,17] a new method to identify tumor-specific antigens (TSAs) was validated,[18] and the role of TSAs as neoantigens was confirmed.[19] Circulating T cells bound to tumor cells were also identified in melanoma patients.[20–23] However, it was only after the human genome sequencing and the development of high-throughput NGS methods that tumor antigens and the T cell receptor (TCR) repertoire in tumors could be characterized on a large scale.

Immune activation and exhaustion

T cell activation requires both TCR engagement and costimulatory signaling.[24,25] Antigen-presenting cells (APCs) present antigenic peptides bound to major histocompatibility complex I or II (MHC) to the TCR complex, which initiates the T cell activation process. T cell activation requires a parallel costimulatory signal to amplify or modulate signals derived from the antigen-bound TCR complex.[26,27] For example, costimulatory molecule CD28 binds to its receptor CD80 (B7-1) or CD86 (B7-2) on APCs to promote T cell activation. These costimulatory signals are regulated by "immune checkpoints" that limit the duration of T cell activation. For instance, checkpoint ligand cytotoxic T lymphocyte antigen 4 (CTLA-4) competes with CD28 for CD80 and CD86. It has a higher affinity than CD28 for these costimulatory molecules, and thus, it interrupts CD28 costimulatory signals. CTLA-4 expressed on the surface of activated T cells blocks IL-2 transcription and inhibits cell cycle progression.[28] Similarly, checkpoint receptor PD-1 (programmed death 1) expressed on the surface of activated T cells binds to its ligands PD-L1 and PD-L2. Once activated, PD-1 directly interferes with TCR signaling and interrupts it. CTLA-4 and PD-1 are two of numerous immune checkpoints expressed by T cells upon activation. Engagement of these checkpoint molecules prevents uncontrolled T cell activation and autoimmunity.[29] Mutations in genes encoding immune checkpoints are associated with autoimmune disorders.[30–32] Prolonged exposure to an antigen can "exhaust" T cells, producing T cell anergy. In tumors and other chronic inflammatory disorders, T cells become anergic, as they are exposed to the same antigens for a prolonged period of time. Anergy is one of many possible mechanisms whereby cancer immune surveillance can fail. Cancer cells also use a variety of local and systemic mechanisms to suppress the immune system. These mechanisms include recruitment of regulatory T cells (T-reg), myeloid-derived suppressor cells that inhibit T cell proliferation, and tumor-associated macrophages that have tumoricidal and tumorostatic functions but in the tumor microenvironment can acquire immune-suppressive, tumor-promoting properties.[33–36] Tumors can also upregulate the PD-L1 checkpoint ligand, which inhibits T cells function through PD-1 on T cells, and produce a host of immunosuppressive factors, such as IL-10, reactive oxygen species, TGF-β, nitric oxide, and others.[37–41]

Cancer immunotherapy

Immune evasion, now established as a fundamental hallmark of cancer, has long been an area of research interest for the development of novel therapeutics in oncology. Immune checkpoints are involved in the maintenance of self-tolerance and modulation of the immune response (as a physiologic negative feedback loop) to prevent autoimmune reactions. However, cancers often hijack immune checkpoints to evade immune destruction. The clinical effectiveness of immune checkpoint inhibitors suggests that this is a common mechanism of evasion.

Checkpoint blockers

A seminal discovery from Dr. Allison's lab showed immune checkpoint CTLA-4 blockade could enhance antitumor immunity in preclinical studies.[42] Monoclonal antibodies (mAbs) blocking human CTLA-4 were developed subsequently, followed by a decade of clinical testing.[43,44] The success of these intense research and clinical trials culminated in FDA approval of first checkpoint inhibitor drug, ipilimumab (Yervoy, Bristol-Myers Squibb, Princeton, NJ) in 2011 for the treatment of patients with metastatic melanoma. Dr. James Allison and Dr. Tasuku Honjo (the discoverer of PD-1) shared the Nobel Prize for Physiology and Medicine in 2018.

The PD-1 pathway provides inhibitory signals to control immune responses. The PD-1 inhibitory receptor regulates T cell activation, effector T cell responses, T cell tolerance, and T cell exhaustion.[45–47] However, PD-1 is expressed on both activated and exhausted T cells, and PD-1 expression alone does not signify T cell exhaustion.[48] PD-1 has two ligands, programmed death ligand PD-L1.[49,50] Tumor cells can express both of these ligands. In the TME, PD-L1 and PD-L2 can be expressed on tumor cells, as well as endothelial cells, stromal cells, APC/myeloid subsets, and T cells, and contribute to the immunosuppressive environment. PD-1 may restrain T cells that have been activated by tumor antigen-bearing dendritic cells (DCs) in lymph nodes or slow down T cells trafficking to the tumor, and effector T cells within the tumor. A tumor cells expressing PD-L1 and/or PD-L2 (or other checkpoint ligands) can hijack the physiological role of immune checkpoints and avoid T cell–mediated elimination. Checkpoint inhibitors block these signals, empowering the immune cells to better identify and attack the cancer cells. Among the numerous immunotherapeutic strategies developed over the past three decades, one of the most significant advances has been the development of mAbs that block immune checkpoint ligands or receptors, thus enhancing and prolonging T cell activation.[51] Pembrolizumab, a PD-1 checkpoint inhibitor, was first approved by the US

FDA for the treatment of melanoma. A number of other mAbs targeting the PD-1/PD-L1 pathway have been approved by FDA, namely, nivolumab (a PD-1 inhibitor, approved in 2014), atezolizumab (a PD-L1 inhibitor, approved in 2016), durvalumab (a PD-L1 inhibitor, approved in 2017), avelumab (PD-L1 inhibitor, approved in 2017), and cemiplimab (a PD-1 inhibitor, approved in 2018). Unlike patient-derived tumor vaccines and other complex biologics, which require patient materials and cells and significant ex vivo manipulations, immune checkpoint mAbs are "off-the-shelf" products that can be used in oncology protocols in a fashion analogous to other antineoplastic mAbs (e.g., rituximab, trastuzumab, pertuzumab, cetuximab, etc.). Immune checkpoint inhibitors have achieved unexpected and remarkable clinical responses in diseases previously deemed incurable, such as non—small-cell lung cancer (NSCLC).[51] That said, even these agents are not "magic bullets." Response rates are in the order of 25%—30%, and resistance develops. Additionally, autoimmune adverse events are common. Therefore, it makes sense to identify patients most likely to respond to these agents, through companion diagnostics that often require immunogenomics. It is no coincidence that the first FDA approval for an antineoplastic agent based on a genomic biomarker rather than tumor site of origin was for pembrolizumab, an anti-PD-1 mAb. In 2017, pembrolizumab was approved for adult and pediatric tumors carrying DNA mismatch repair (dMMR) defects or high microsatellite instability (MSI-high), also an indication of genomic instability.[52] Tumors carrying these genomic markers were highly sensitive to pembrolizumab, whereas tumors not carrying these markers were not, irrespective of anatomical site of origin. It is likely, though not formally proven in the registration trial, that genomically unstable tumors have a higher tumor antigen load and are therefore more immunogenic. This FDA approval officially opened the era of clinical immunogenomics. Table 7.1 offers a list of completed clinical trials with checkpoints inhibitor as single agents in various cancers. Numerous other clinical trials are underway using checkpoint inhibitors either as a single agent or in combination with other therapeutic agents (https://clinicaltrials.gov/).

Combinatorial approaches

A combinatorial immune checkpoint-blocking approach to treat diseases not only might increase the percentage of treatment responders but could also broaden the spectrum of malignancies for which immunotherapy may be effective. Matsuzaki et al.[53] demonstrated that human tumor-derived CD8+ T cells express both PD-1 and another immune checkpoint molecule, lymphocyte activation gene-3 (LAG-3). Abrogating the effects of either molecule alone was not sufficient to restore T cell functions in vitro. Blocking PD-1 and LAG-3 together with their respective mAbs successfully bolstered T cell proliferation and cytokine production, consistent with findings from animal tumor models.[54] Combinatorial strategies have also demonstrated success in the clinic. Wolchok et al.[55] demonstrated a 40% objective response rate in a cohort of 52 patients with melanoma who received anti-PD-1 and anti-CTLA4 concurrently.

Table 7.1 Completed clinical trials using check point inhibitors as a single agent (Clinicaltrials.gov)

Interventions	Completed Study Title	Conditions	ClinicalTrials.gov Identifier
PD-1	Study of Pembrolizumab (MK-3475) in Participants With Progressive Locally Advanced or Metastatic Carcinoma, Melanoma, or Non-small Cell Lung Carcinoma (P07990/ MK-3475-001/KEYNOTE-001) (KEYNOTE-001)	Cancer, Solid Tumor	NCT01295827
	Anti-PD-1 Monoclonal Antibody in Advanced, Trastuzumab-resistant, HER2-positive Breast Cancer (PANACEA)	Metastatic Breast Cancer	NCT02129556
	ONO-4538 Phase I Study in Patients With Advanced Malignant Solid Tumors in Japan	Malignant Solid Tumor	NCT00836888
PDL-1	A Study of Atezolizumab (an Engineered Anti-Programmed Death-Ligand 1 [PDL1] Antibody) to Evaluate Safety, Tolerability and Pharmacokinetics in Participants With Locally Advanced or Metastatic Solid Tumors	Tumors, Hematologic Malignancies	NCT01375842
	A Study of Atezolizumab in Participants With Programmed Death - Ligand 1 (PD-L1) Positive Locally Advanced or Metastatic Non-Small Cell Lung Cancer (BIRCH)	Non-Small Cell Lung Cancer	NCT02031458
	A Study of Atezolizumab in Participants With Programmed Death-Ligand 1 (PD-L1) Positive Locally Advanced or Metastatic Non-Small Cell Lung Cancer (NSCLC) [FIR]	Non-Small Cell Lung Cancer	NCT01846416
CTLA-4	MDX-010 in Treating Patients With Stage IV Pancreatic Cancer That Cannot Be Removed By Surgery	Pancreatic Cancer	NCT00112580

Table 7.1 Completed clinical trials using check point inhibitors as a single agent (Clinicaltrials.gov)—*cont'd*

Interventions	Completed Study Title	Conditions	ClinicalTrials.gov Identifier
	Study to Compare Two Formulations of CP-675,206 Monoclonal Antibody	Melanoma	NCT00431275

Chimeric antigen receptor T cell therapy

In 2017, the approval of Kymirah, a chimeric antigen receptor (CAR) T cell therapy, by the US FDA, was momentous. Two more therapies using genetically engineered T cells against lymphoma or leukemia have won US FDA approval since then, and hundreds of clinical trials are presently ongoing leveraging CAR T cell therapies potential for other malignancies, including solid tumors.[56–59] CAR T cell therapy, in essence, is a "living drug" derived from the patient's own blood. In the CAR T cell manufacturing process, T cells are purified from the blood of the patients before being genetically engineered to express a CAR. This CAR molecule is armored with an extracellular antibody-like domain fused to an intracellular signaling domain, which enables T cells to simultaneously detect and kill cancer cells and while boosting T cell activity. These engineered T cells are then expanded in sufficient number ex vivo before being infused back into the patient. A successful treatment is supposed to see CAR T cell further expand in the body while homing in on and killing target cancer cells. Although CAR T cell technology is 30 years old, understanding its therapeutic potential for solid tumors remains a work in progress. Additionally, CAR T cells do have potentially serious adverse effects including cytokine release syndrome (CRS). That said, development efforts are underway for new generations of safer CAR-T cells that include sophisticated genetic circuitry to control activation.[60,61] Third-generation CAR receptors all have common components including an extracellular targeting element, transmembrane spacer, two costimulatory domains, and a signaling domain.[62–64] They are ideally specific for tumor neoantigens and capable of recognizing intact membrane antigens in an MHC-unrestricted fashion. Bypassing the steps of antigen processing and presentation by MHC helps alleviate the risk of immune escape due to loss of MHC expression. The combination of a stimulatory and costimulatory domain tethered to one receptor improves proliferative responses following antigen binding and prevents T cell anergy or exhaustion, leading to durable T cell response. The ultimate goal for CAR-T therapy would be the development of an "off-the-shelf" cell product that does not require ad hoc genetic manipulation of each patient's T cells. Such a product could greatly expand the usefulness and improve the cost-effectiveness of CAR T cell therapy.

Oncolytic virus therapy

Oncolytic virotherapy uses genetically engineered, replication-incompetent viruses to directly eliminate cancer cells and stimulate antitumor immunity. Talimogene laherparepvec (T-VEC) is the first approved oncolytic virus immunotherapy for late-stage melanoma patients by the US FDA.[65–67] T-VEC is a recombinant, attenuated herpes simplex-1 virus. T-VEC is also being tested for other cancers, either as single agent or in combination with standard therapies. Recombinant polioviruses are also being tested as cancer virotherapeutics for malignancies expressing the poliovirus receptor, CD155.[68,69]

Cancer vaccines

Cancer vaccines have been investigated for decades. Vaccines against multiple serotypes of human papillomavirus (HPV) are effective in preventing HPV-associated cancers (e.g., cervical, oropharyngeal, penile, anal carcinomas).[70–72] DC-based vaccines have shown reasonable success so far in treating cancer.[73–76] Sipuleucel-T (Provenge), approved for treating advanced prostate cancer, is an example of a DC vaccine. However, DC vaccines suffer from the same practical shortcomings as CAR-T cells, in which they must be manufactured from the person in whom they will be used. The process is cost-, labor-, and time-intensive. Research is underway to accelerate turnaround times and improve cost-effectiveness. A clinical trial is currently exploring the safety and efficacy of a DC vaccine in metastatic melanoma.[77] Synthetic peptide vaccines can be used in principle to elicit and expand tumor-specific T cells capable of controlling or eradicating cancers. However, despite early success in preclinical studies, clinical trials with peptide vaccine have not been successful to date. Recent progress in understanding the critical roles of immune adjuvants as well as improvements in formulation and modifications of immunogenicity made possible by important basic discoveries have led to a renewed interest in the use of peptide vaccines in oncology. A promising trial found that a peptide-based vaccine targeting indoleamine 2,3-dioxygenase (IDO1) in combination with standard-of-care chemotherapy-prolonged progression-free survival in nearly 50% of patients with NSCLC.[78]

Researchers have equipped innate immune system cells, such as natural killer cells (NK cells), macrophages with CAR, and CAR-NK, cells have made their debut in clinical trials. The first clinical trial started in China in 2016 and preliminary data indicate safety and tolerability.[79] A CAR-NK "off-the-shelf" product may be easier to develop than a T cell—based product.

The technologies of immunogenomics

The development of NGS technologies over the past decade facilitated our ability to identify tumor-specific neoantigens in a rapid and comprehensive manner.[13,80]

Concomitant development of computational algorithms has also enabled researchers to identify all classes of somatic variants (nonsynonymous point mutations, insertions, deletions, and copy number variants) from NGS data. Advances in NGS-based exome sequencing (e.g., depth and uniformity of coverage across genomic sequences) of tumor samples and control germline DNA have produced tremendous insights into the process of cancer evolution and the development of therapy resistance in advanced cancers. It is now well established that cancers are often multiclonal and that clones characterized by different mutational profiles have different sensitivity to standard-of-care therapies (and presumably immunotherapies). Thus, cancer clones undergo a process of quasi-Darwinian evolution that selects treatment-resistant clones.[81] A consequence of this is that the mutational profiles of cancers evolve in time. Bioinformatic algorithms can now identify "founder" clones versus subclonal mutations in NGS data sets and create "tropical fish plots" and other visualization tools that graphically describe the clonal evolution of cancers.[81] Founder mutations are defined as the original mutations that occurred during the process of neoplastic transformation, and subclonal mutations occur during tumor progression.[13] It is important to understand that neoantigens resulting from founder clone mutations could potentially generate T cell response to all tumor cells, but neoantigens from subclonal mutation would elicit a T cell response to only specific clones carrying that mutation. If immunogenic mutations are not indispensable for the survival of cancer cells, they can be lost through clonal selection under therapy. This is a likely mechanism of acquired resistance.

It is critical to use appropriate depth coverage and algorithm(s) to identify antigenic variant(s) from NGS data. Single-nucleotide variants (point mutations) are easy to identify using standard alignment algorithms to the reference human genome. But "indels" (insertion or deletion of one or more nucleotides) are difficult to identify using standard alignment algorithms. Importantly, the advancement of computational approaches such as gapped alignment, split-read algorithms, or assembly-based realignment helps significantly to improve the quality of "indel" detection.[13,82,83] Another hurdle to the identification of antigenic somatic variations are neoantigenic sequences resulting from the fusion of two protein-coding sequences or from aberrant RNA splicing. DNA exome sequencing is not appropriate to identify these variants, but RNA-based analysis (RNA sequencing) brings added value to identify these fusion transcripts.[84–87] Zhang et al. described a process called INTEGRATE-Neo to identify fusion peptide neoantigens in prostate cancer.[88] Importantly, RNA sequencing requires reasonably undegraded RNA. Unlike DNA, RNA is extremely labile under conditions commonly encountered in operating rooms and pathology grossing labs due to ribonuclease contamination and inherent chemical instability. While protocols exist for RNA extraction and sequencing even from formalin-fixed paraffin-embedded tissue (FFPE), preanalytic variability in RNA quality must be addressed to assure reliable results. Following variant detection, various computational tools (such as Annovar and VEP) are utilized to generate translated peptide to predict neoantigenic properties. Next, neoantigen prediction software is used to predict both class I and class II HLA binding affinities for each mutant peptide. A detailed workflow for neoantigen discovery and personalized cancer vaccine design was described by Mardis et al.[13] The Immune Epitope Database (IEDB) (http://tools.immuneepitope.org/main/) offers many binding prediction software as well as immunogenomic algorithms.[89] In addition, publicly available database such as pVAC-seq (https://github.com/griffithlab/pVAC-Seq) and epidisco (https://github.com/hammerlab) can also be used for neoantigen predictions.[13]

T cell receptor and B cell receptor sequencing

TCRs and B cell receptors (BCRs) recognize specific antigens and selectively activate T cells and B cells, respectively, during adaptive immune system activation.[90] During the process of lymphocyte development and differentiation, genes encoding TCRs and BCRs experience a complex rearrangement (homologous recombination) to produce a broad repertoire of functionally diverse receptors. Different T and B cell populations reactive to different antigens will have different TCR or BCR variable region sequences. NGS has enabled us to analyze and understand the genetic diversity of TCR and BCR repertoires.[91–94] In the era of immunotherapy, it is very important to identify neoantigen-specific T cell and B cell populations in cancer patients. This can be used to predict responses to immunotherapy or to identify candidate antigens for vaccines or for CAR T cell adoptive therapy. Recently, Nakamura et al.[95] described a detailed protocol to sequence TCR and BCR genes using NGS. These authors were able to generate 10 million sequence reads of the receptors genes and complete the characterization of TCRs or BCRs repertories.[95] Using their protocol, they identified tumor-specific infiltrating T cell populations from solid tumors or malignant ascites.[96,97] The same group applied this approach to identify the progression of specific T cell populations during the course of treatment.[95,98–106] Nakamura et al. also described a TCR sequencing protocol to identify neoantigen-specific TCR-engineered T cells using patient's blood.[107] TCR sequencing also provides a comprehensive characterization of T cell population dynamics during immunotherapy.[95,100]

Role of precision medicine in immunogenomics

The most polymorphic region in the human genome is found in chromosome 6q21.3.[108] This region contains MHC genes and numerous other genes associated with immunity and autoimmune disorders. A recent effort to map the architecture of polymorphisms in the human genome confirmed that genes controlling immune responses are among the most variable.[109] This strongly suggests significant genetic variability in immune responses and most likely individual variability in the safety and efficacy of immunotherapeutics. Population genomics will play a critical role in identifying genetic markers predictive of immunological outcomes. The "precision medicine initiative" was first announced by US president Barack Obama in his 2015

State of the Union address. It is currently taking the form of a massive project, "**The All of Us**" Research Program (https://allofus.nih.gov/).[110] "All of Us" is an NIH-funded, historic research effort to create a comprehensive, high-quality database including clinical and genomic data from a million or more individuals covering every ethnicity, including groups traditionally underrepresented in medical research. At the time of this writing, the program has enrolled more than 200,000 participants. The formidable data set resulting from this program, once fully assembled, can reveal fundamental mechanistic insights into human disease. Information on the highly polymorphic genes controlling immune responses will eventually allow us to elucidate the role of germline genetics in every patient's propensity to respond to immunotherapeutics and/or suffer from immune-mediated adverse events.

Conclusions

The widespread availability and decreasing cost of NGS-based genomic technologies have generated novel mechanistic insights into the genetic control of immune responses, the genomic landscape and clonal architecture of human tumors, and the complexity of tumor-immune system interactions. These insights are rapidly being translated to the clinic, with new generations of genetically engineered immunotherapeutics and companion diagnostics based on genomic platforms. We can now envision a future when the patient's germline genome and tumor mutational and gene expression profiles are used in routine clinical practice to guide immunotherapeutic decisions.

References

1. Schreiber RD, Old LJ, Smyth MJ. Cancer immunoediting: integrating immunity's roles in cancer suppression and promotion. *Science* 2011;**331**:1565−70.
2. Chen DS, Mellman I. Elements of cancer immunity and the cancer-immune set point. *Nature* 2017;**541**:321−30.
3. Matsushita H, Vesely MD, Koboldt DC, Rickert CG, Uppaluri R, Magrini VJ, Arthur CD, White JM, Chen YS, Shea LK, Hundal J, Wendl MC, Demeter R, Wylie T, Allison JP, Smyth MJ, Old LJ, Mardis ER, Schreiber RD. Cancer exome analysis reveals a T-cell-dependent mechanism of cancer immunoediting. *Nature* 2012;**482**: 400−4.
4. Rizvi NA, Hellmann MD, Snyder A, Kvistborg P, Makarov V, Havel JJ, Lee W, Yuan J, Wong P, Ho TS, Miller ML, Rekhtman N, Moreira AL, Ibrahim F, Bruggeman C, Gasmi B, Zappasodi R, Maeda Y, Sander C, Garon EB, Merghoub T, Wolchok JD, Schumacher TN, Chan TA. Cancer immunology. Mutational landscape determines sensitivity to PD-1 blockade in non-small cell lung cancer. *Science* 2015;**348**:124−8.
5. Robbins PF, Lu YC, El-Gamil M, Li YF, Gross C, Gartner J, Lin JC, Teer JK, Cliften P, Tycksen E, Samuels Y, Rosenberg SA. Mining exomic sequencing data to identify

mutated antigens recognized by adoptively transferred tumor-reactive T cells. *Nat Med* 2013;**19**:747–52.

6. Sharma P, Allison JP. The future of immune checkpoint therapy. *Science* 2015;**348**: 56–61.
7. Snyder A, Makarov V, Merghoub T, Yuan J, Zaretsky JM, Desrichard A, Walsh LA, Postow MA, Wong P, Ho TS, Hollmann TJ, Bruggeman C, Kannan K, Li Y, Elipenahli C, Liu C, Harbison CT, Wang L, Ribas A, Wolchok JD, Chan TA. Genetic basis for clinical response to CTLA-4 blockade in melanoma. *N Engl J Med* 2014; **371**:2189–99.
8. Tran E, Turcotte S, Gros A, Robbins PF, Lu YC, Dudley ME, Wunderlich JR, Somerville RP, Hogan K, Hinrichs CS, Parkhurst MR, Yang JC, Rosenberg SA. Cancer immunotherapy based on mutation-specific CD4+ T cells in a patient with epithelial cancer. *Science* 2014;**344**:641–5.
9. van Rooij N, van Buuren MM, Philips D, Velds A, Toebes M, Heemskerk B, van Dijk LJ, Behjati S, Hilkmann H, El Atmioui D, Nieuwland M, Stratton MR, Kerkhoven RM, Kesmir C, Haanen JB, Kvistborg P, Schumacher TN. Tumor exome analysis reveals neoantigen-specific T-cell reactivity in an ipilimumab-responsive melanoma. *J Clin Oncol* 2013;**31**:e439–442.
10. Foley EJ. Antigenic properties of methylcholanthrene-induced tumors in mice of the strain of origin. *Cancer Res* 1953;**13**:835–7.
11. Old LJ, Boyse EA. Immunology of experimental tumors. *Annu Rev Med* 1964;**15**: 167–86.
12. Prehn RT, Main JM. Immunity to methylcholanthrene-induced sarcomas. *J Natl Cancer Inst* 1957;**18**:769–78.
13. Liu XS, Mardis ER. Applications of immunogenomics to cancer. *Cell* 2017;**168**: 600–12.
14. Babbitt BP, Allen PM, Matsueda G, Haber E, Unanue ER. Binding of immunogenic peptides to Ia histocompatibility molecules. *Nature* 1985;**317**:359–61.
15. Bjorkman PJ, Saper MA, Samraoui B, Bennett WS, Strominger JL, Wiley DC. Structure of the human class I histocompatibility antigen, HLA-A2. *Nature* 1987;**329**:506–12.
16. Cerottini JC, Engers HD, Macdonald HR, Brunner T. Generation of cytotoxic T lymphocytes in vitro. I. Response of normal and immune mouse spleen cells in mixed leukocyte cultures. *J Exp Med* 1974;**140**:703–17.
17. Gillis S, Smith KA. Long term culture of tumour-specific cytotoxic T cells. *Nature* 1977;**268**:154–6.
18. De Plaen E, Lurquin C, Van Pel A, Mariame B, Szikora JP, Wolfel T, Sibille C, Chomez P, Boon T. Immunogenic (tum-) variants of mouse tumor P815: cloning of the gene of tum- antigen P91A and identification of the tum- mutation. *Proc Natl Acad Sci USA* 1988;**85**:2274–8.
19. Monach PA, Meredith SC, Siegel CT, Schreiber H. A unique tumor antigen produced by a single amino acid substitution. *Immunity* 1995;**2**:45–59.
20. Knuth A, Danowski B, Oettgen HF, Old LJ. T-cell-mediated cytotoxicity against autologous malignant melanoma: analysis with interleukin 2-dependent T-cell cultures. *Proc Natl Acad Sci USA* 1984;**81**:3511–5.
21. Van den Eynde B, Hainaut P, Herin M, Knuth A, Lemoine C, Weynants P, van der Bruggen P, Fauchet R, Boon T. Presence on a human melanoma of multiple antigens recognized by autologous CTL. *Int J Cancer* 1989;**44**:634–40.

22. Robbins PF, El-Gamil M, Li YF, Kawakami Y, Loftus D, Appella E, Rosenberg SA. A mutated beta-catenin gene encodes a melanoma-specific antigen recognized by tumor infiltrating lymphocytes. *J Exp Med* 1996;**183**:1185–92.
23. Dubey P, Hendrickson RC, Meredith SC, Siegel CT, Shabanowitz J, Skipper JC, Engelhard VH, Hunt DF, Schreiber H. The immunodominant antigen of an ultraviolet-induced regressor tumor is generated by a somatic point mutation in the DEAD box helicase p68. *J Exp Med* 1997;**185**:695–705.
24. Bretscher P, Cohn M. A theory of self-nonself discrimination. *Science* 1970;**169**: 1042–9.
25. Lafferty KJ, Cunningham AJ. A new analysis of allogeneic interactions. *Aust J Exp Biol Med Sci* 1975;**53**:27–42.
26. Sharpe AH, Freeman GJ. The B7-CD28 superfamily. *Nat Rev Immunol* 2002;**2**:116–26.
27. Kroczek RA, Mages HW, Hutloff A. Emerging paradigms of T-cell co-stimulation. *Curr Opin Immunol* 2004;**16**:321–7.
28. Brunner MC, Chambers CA, Chan FK, Hanke J, Winoto A, Allison JP. CTLA-4-Mediated inhibition of early events of T cell proliferation. *J Immunol* 1999;**162**: 5813–20.
29. Page DB, Postow MA, Callahan MK, Allison JP, Wolchok JD. Immune modulation in cancer with antibodies. *Annu Rev Med* 2014;**65**:185–202.
30. Schreiner B, Mitsdoerffer M, Kieseier BC, Chen L, Hartung HP, Weller M, Wiendl H. Interferon-beta enhances monocyte and dendritic cell expression of B7-H1 (PD-L1), a strong inhibitor of autologous T-cell activation: relevance for the immune modulatory effect in multiple sclerosis. *J Neuroimmunol* 2004;**155**:172–82.
31. Marson A, Housley WJ, Hafler DA. Genetic basis of autoimmunity. *J Clin Investig* 2015;**125**:2234–41.
32. Paluch C, Santos AM, Anzilotti C, Cornall RJ, Davis SJ. Immune checkpoints as therapeutic targets in autoimmunity. *Front Immunol* 2018;**9**:2306.
33. Gabrilovich DI, Bronte V, Chen SH, Colombo MP, Ochoa A, Ostrand-Rosenberg S, Schreiber H. The terminology issue for myeloid-derived suppressor cells. *Cancer Res* 2007;**67**:425. author reply 426.
34. Ugel S, De Sanctis F, Mandruzzato S, Bronte V. Tumor-induced myeloid deviation: when myeloid-derived suppressor cells meet tumor-associated macrophages. *J Clin Investig* 2015;**125**:3365–76.
35. Bronte V, Brandau S, Chen SH, Colombo MP, Frey AB, Greten TF, Mandruzzato S, Murray PJ, Ochoa A, Ostrand-Rosenberg S, Rodriguez PC, Sica A, Umansky V, Vonderheide RH, Gabrilovich DI. Recommendations for myeloid-derived suppressor cell nomenclature and characterization standards. *Nat Commun* 2016;**7**:12150.
36. Kumar V, Patel S, Tcyganov E, Gabrilovich DI. The nature of myeloid-derived suppressor cells in the tumor microenvironment. *Trends Immunol* 2016;**37**:208–20.
37. Fong L, Small EJ. Anti-cytotoxic T-lymphocyte antigen-4 antibody: the first in an emerging class of immunomodulatory antibodies for cancer treatment. *J Clin Oncol* 2008;**26**:5275–83.
38. Bronte V, Mocellin S. Suppressive influences in the immune response to cancer. *J Immunother* 2009;**32**:1–11.
39. Poschke I, Mougiakakos D, Kiessling R. Camouflage and sabotage: tumor escape from the immune system. *Cancer Immunol Immunother* 2011;**60**:1161–71.
40. Mocellin S, Nitti D. CTLA-4 blockade and the renaissance of cancer immunotherapy. *Biochim Biophys Acta* 2013;**1836**:187–96.

41. Buchbinder EI, McDermott DF. Cytotoxic T-lymphocyte antigen-4 blockade in melanoma. *Clin Ther* 2015;**37**:755–63.
42. Leach DR, Krummel MF, Allison JP. Enhancement of antitumor immunity by CTLA-4 blockade. *Science* 1996;**271**:1734–6.
43. Hodi FS, O'Day SJ, McDermott DF, Weber RW, Sosman JA, Haanen JB, Gonzalez R, Robert C, Schadendorf D, Hassel JC, Akerley W, van den Eertwegh AJ, Lutzky J, Lorigan P, Vaubel JM, Linette GP, Hogg D, Ottensmeier CH, Lebbe C, Peschel C, Quirt I, Clark JI, Wolchok JD, Weber JS, Tian J, Yellin MJ, Nichol GM, Hoos A, Urba WJ. Improved survival with ipilimumab in patients with metastatic melanoma. *N Engl J Med* 2010;**363**:711–23.
44. Robert C, Thomas L, Bondarenko I, O'Day S, Weber J, Garbe C, Lebbe C, Baurain JF, Testori A, Grob JJ, Davidson N, Richards J, Maio M, Hauschild A, Miller Jr WH, Gascon P, Lotem M, Harmankaya K, Ibrahim R, Francis S, Chen TT, Humphrey R, Hoos A, Wolchok JD. Ipilimumab plus dacarbazine for previously untreated metastatic melanoma. *N Engl J Med* 2011;**364**:2517–26.
45. Barber DL, Wherry EJ, Masopust D, Zhu B, Allison JP, Sharpe AH, Freeman GJ, Ahmed R. Restoring function in exhausted CD8 T cells during chronic viral infection. *Nature* 2006;**439**:682–7.
46. Francisco LM, Sage PT, Sharpe AH. The PD-1 pathway in tolerance and autoimmunity. *Immunol Rev* 2010;**236**:219–42.
47. LaFleur MW, Muroyama Y, Drake CG, Sharpe AH. Inhibitors of the PD-1 pathway in tumor therapy. *J Immunol* 2018;**200**:375–83.
48. Schildberg FA, Klein SR, Freeman GJ, Sharpe AH. Coinhibitory pathways in the B7-CD28 ligand-receptor family. *Immunity* 2016;**44**:955–72.
49. Freeman GJ, Long AJ, Iwai Y, Bourque K, Chernova T, Nishimura H, Fitz LJ, Malenkovich N, Okazaki T, Byrne MC, Horton HF, Fouser L, Carter L, Ling V, Bowman MR, Carreno BM, Collins M, Wood CR, Honjo T. Engagement of the PD-1 immunoinhibitory receptor by a novel B7 family member leads to negative regulation of lymphocyte activation. *J Exp Med* 2000;**192**:1027–34.
50. Latchman Y, Wood CR, Chernova T, Chaudhary D, Borde M, Chernova I, Iwai Y, Long AJ, Brown JA, Nunes R, Greenfield EA, Bourque K, Boussiotis VA, Carter LL, Carreno BM, Malenkovich N, Nishimura H, Okazaki T, Honjo T, Sharpe AH, Freeman GJ. PD-L2 is a second ligand for PD-1 and inhibits T cell activation. *Nat Immunol* 2001;**2**:261–8.
51. Haanen JB, Robert C. Immune checkpoint inhibitors. *Prog Tumor Res* 2015;**42**:55–66.
52. Marcus L, Lemery SJ, Keegan P, Pazdur R. FDA approval summary: pembrolizumab for the treatment of microsatellite instability-high solid tumors. *Clin Cancer Res* 2019;**25**:3753–8.
53. Matsuzaki J, Gnjatic S, Mhawech-Fauceglia P, Beck A, Miller A, Tsuji T, Eppolito C, Qian F, Lele S, Shrikant P, Old LJ, Odunsi K. Tumor-infiltrating NY-ESO-1-specific CD8+ T cells are negatively regulated by LAG-3 and PD-1 in human ovarian cancer. *Proc Natl Acad Sci USA* 2010;**107**:7875–80.
54. Woo SR, Turnis ME, Goldberg MV, Bankoti J, Selby M, Nirschl CJ, Bettini ML, Gravano DM, Vogel P, Liu CL, Tangsombatvisit S, Grosso JF, Netto G, Smeltzer MP, Chaux A, Utz PJ, Workman CJ, Pardoll DM, Korman AJ, Drake CG, Vignali DA. Immune inhibitory molecules LAG-3 and PD-1 synergistically regulate T-cell function to promote tumoral immune escape. *Cancer Res* 2012;**72**:917–27.

55. Wolchok JD, Kluger H, Callahan MK, Postow MA, Rizvi NA, Lesokhin AM, Segal NH, Ariyan CE, Gordon RA, Reed K, Burke MM, Caldwell A, Kronenberg SA, Agunwamba BU, Zhang X, Lowy I, Inzunza HD, Feely W, Horak CE, Hong Q, Korman AJ, Wigginton JM, Gupta A, Sznol M. Nivolumab plus ipilimumab in advanced melanoma. *N Engl J Med* 2013;**369**:122–33.
56. Ma L, Dichwalkar T, Chang JYH, Cossette B, Garafola D, Zhang AQ, Fichter M, Wang C, Liang S, Silva M, Kumari S, Mehta NK, Abraham W, Thai N, Li N, Wittrup KD, Irvine DJ. Enhanced CAR-T cell activity against solid tumors by vaccine boosting through the chimeric receptor. *Science* 2019;**365**:162–8.
57. Khalil DN, Smith EL, Brentjens RJ, Wolchok JD. The future of cancer treatment: immunomodulation, CARs and combination immunotherapy. *Nat Rev Clin Oncol* 2016;**13**:394.
58. Newick K, O'Brien S, Moon E, Albelda SM. CAR T cell therapy for solid tumors. *Annu Rev Med* 2017;**68**:139–52.
59. Lamprecht M, Dansereau C. CAR T-cell therapy: update on the state of the science. *Clin J Oncol Nurs* 2019;**23**:6–12.
60. Caliendo F, Dukhinova M, Siciliano V. Engineered cell-based therapeutics: synthetic biology meets immunology. *Front Bioeng Biotechnol* 2019;**7**:43.
61. DeRenzo C, Gottschalk S. Genetic modification strategies to enhance CAR T cell persistence for patients with solid tumors. *Front Immunol* 2019;**10**:218.
62. Tang XY, Sun Y, Zhang A, Hu GL, Cao W, Wang DH, Zhang B, Chen H. Third-generation CD28/4-1BB chimeric antigen receptor T cells for chemotherapy relapsed or refractory acute lymphoblastic leukaemia: a non-randomised, open-label phase I trial protocol. *BMJ Open* 2016;**6**:e013904.
63. Van Schandevyl S, Kerre T. Chimeric antigen receptor T-cell therapy: design improvements and therapeutic strategies in cancer treatment. *Acta Clin Belg* 2018:1–7.
64. Zhang C, Liu J, Zhong JF, Zhang X. Engineering CAR-T cells. *Biomark Res* 2017;**5**:22.
65. Johnson DB, Puzanov I, Kelley MC. Talimogene laherparepvec (T-VEC) for the treatment of advanced melanoma. *Immunotherapy* 2015;**7**:611–9.
66. Kohlhapp FJ, Zloza A, Kaufman HL. Talimogene laherparepvec (T-VEC) as cancer immunotherapy. *Drugs Today* 2015;**51**:549–58.
67. Bommareddy PK, Patel A, Hossain S, Kaufman HL. Talimogene laherparepvec (T-VEC) and other oncolytic viruses for the treatment of melanoma. *Am J Clin Dermatol* 2017;**18**:1–15.
68. Brown MC, Holl EK, Boczkowski D, Dobrikova E, Mosaheb M, Chandramohan V, Bigner DD, Gromeier M, Nair SK. Cancer immunotherapy with recombinant poliovirus induces IFN-dominant activation of dendritic cells and tumor antigen-specific CTLs. *Sci Transl Med* 2017;**9**.
69. Peruzzi P, Chiocca EA. Viruses in cancer therapy - from benchwarmers to quarterbacks. *Nat Rev Clin Oncol* 2018;**15**:657–8.
70. Hancock G, Hellner K, Dorrell L. Therapeutic HPV vaccines. *Best Pract Res Clin Obstet Gynaecol* 2018;**47**:59–72.
71. Schellenbacher C, Roden RBS, Kirnbauer R. Developments in L2-based human papillomavirus (HPV) vaccines. *Virus Res* 2017;**231**:166–75.
72. World Health Organization, Electronic address, s. w. i. Human papillomavirus vaccines: WHO position paper, May 2017-Recommendations. *Vaccine* 2017;**35**:5753–5.
73. Gilboa E. DC-based cancer vaccines. *J Clin Investig* 2007;**117**:1195–203.

74. Palucka K, Banchereau J. Dendritic-cell-based therapeutic cancer vaccines. *Immunity* 2013;**39**:38−48.
75. Santos PM, Butterfield LH. Dendritic cell-based cancer vaccines. *J Immunol* 2018;**200**: 443−9.
76. Saxena M, Bhardwaj N. Re-emergence of dendritic cell vaccines for cancer treatment. *Trends Cancer* 2018;**4**:119−37.
77. Dillman RO, Cornforth AN, Nistor GI, McClay EF, Amatruda TT, Depriest C. Randomized phase II trial of autologous dendritic cell vaccines versus autologous tumor cell vaccines in metastatic melanoma: 5-year follow up and additional analyses. *J Immunother Cancer* 2018;**6**:19.
78. Iversen TZ, Engell-Noerregaard L, Ellebaek E, Andersen R, Larsen SK, Bjoern J, Zeyher C, Gouttefangeas C, Thomsen BM, Holm B, Thor Straten P, Mellemgaard A, Andersen MH, Svane IM. Long-lasting disease stabilization in the absence of toxicity in metastatic lung cancer patients vaccinated with an epitope derived from indoleamine 2,3 dioxygenase. *Clin Cancer Res* 2014;**20**:221−32.
79. Leslie M. New cancer-fighting cells enter trials. *Science* 2018;**361**:1056−7.
80. Gubin MM, Artyomov MN, Mardis ER, Schreiber RD. Tumor neoantigens: building a framework for personalized cancer immunotherapy. *J Clin Investig* 2015;**125**:3413−21.
81. Krzywinski M. Visualizing clonal evolution in cancer. *Mol Cell* 2016;**62**:652−6.
82. Mose LE, Wilkerson MD, Hayes DN, Perou CM, Parker JS. ABRA: improved coding indel detection via assembly-based realignment. *Bioinformatics* 2014;**30**:2813−5.
83. Narzisi G, O'Rawe JA, Iossifov I, Fang H, Lee YH, Wang Z, Wu Y, Lyon GJ, Wigler M, Schatz MC. Accurate de novo and transmitted indel detection in exome-capture data using microassembly. *Nat Methods* 2014;**11**:1033−6.
84. Li Y, Chien J, Smith DI, Ma J. FusionHunter: identifying fusion transcripts in cancer using paired-end RNA-seq. *Bioinformatics* 2011;**27**:1708−10.
85. Scolnick JA, Dimon M, Wang IC, Huelga SC, Amorese DA. An efficient method for identifying gene fusions by targeted RNA sequencing from fresh frozen and FFPE samples. *PLoS One* 2015;**10**:e0128916.
86. Kumar S, Razzaq SK, Vo AD, Gautam M, Li H. Identifying fusion transcripts using next generation sequencing. *Wiley Interdiscip Rev RNA* 2016;**7**:811−23.
87. Zhang J, White NM, Schmidt HK, Fulton RS, Tomlinson C, Warren WC, Wilson RK, Maher CA. INTEGRATE: gene fusion discovery using whole genome and transcriptome data. *Genome Res* 2016;**26**:108−18.
88. Zhang J, Mardis ER, Maher CA. INTEGRATE-neo: a pipeline for personalized gene fusion neoantigen discovery. *Bioinformatics* 2017;**33**:555−7.
89. Robinson J, Halliwell JA, McWilliam H, Lopez R, Parham P, Marsh SG. The IMGT/HLA database. *Nucleic Acids Res* 2013;**41**:D1222−7.
90. Borst J, Jacobs H, Brouns G. Composition and function of T-cell receptor and B-cell receptor complexes on precursor lymphocytes. *Curr Opin Immunol* 1996;**8**:181−90.
91. Bashford-Rogers RJ, Palser AL, Huntly BJ, Rance R, Vassiliou GS, Follows GA, Kellam P. Network properties derived from deep sequencing of human B-cell receptor repertoires delineate B-cell populations. *Genome Res* 2013;**23**:1874−84.
92. Boyd SD, Gaeta BA, Jackson KJ, Fire AZ, Marshall EL, Merker JD, Maniar JM, Zhang LN, Sahaf B, Jones CD, Simen BB, Hanczaruk B, Nguyen KD, Nadeau KC, Egholm M, Miklos DB, Zehnder JL, Collins AM. Individual variation in the germline Ig gene repertoire inferred from variable region gene rearrangements. *J Immunol* 2010; **184**:6986−92.

93. Freeman JD, Warren RL, Webb JR, Nelson BH, Holt RA. Profiling the T-cell receptor beta-chain repertoire by massively parallel sequencing. *Genome Res* 2009;**19**:1817–24.
94. Lange V, Bohme I, Hofmann J, Lang K, Sauter J, Schone B, Paul P, Albrecht V, Andreas JM, Baier DM, Nething J, Ehninger U, Schwarzelt C, Pingel J, Ehninger G, Schmidt AH. Cost-efficient high-throughput HLA typing by MiSeq amplicon sequencing. *BMC Genomics* 2014;**15**:63.
95. Zewde M, Kiyotani K, Park JH, Fang H, Yap KL, Yew PY, Alachkar H, Kato T, Mai TH, Ikeda Y, Matsuda T, Liu X, Ren L, Deng B, Harada M, Nakamura Y. The era of immunogenomics/immunopharmacogenomics. *J Hum Genet* 2018;**63**:865–75.
96. Jang M, Yew PY, Hasegawa K, Ikeda Y, Fujiwara K, Fleming GF, Nakamura Y, Park JH. Characterization of T cell repertoire of blood, tumor, and ascites in ovarian cancer patients using next generation sequencing. *Oncoimmunology* 2015;**4**:e1030561.
97. Liu X, Venkataraman G, Lin J, Kiyotani K, Smith S, Montoya M, Nakamura Y, Kline J. Highly clonal regulatory T-cell population in follicular lymphoma - inverse correlation with the diversity of CD8(+) T cells. *Oncoimmunology* 2015;**4**:e1002728.
98. Yew PY, Alachkar H, Yamaguchi R, Kiyotani K, Fang H, Yap KL, Liu HT, Wickrema A, Artz A, van Besien K, Imoto S, Miyano S, Bishop MR, Stock W, Nakamura Y. Quantitative characterization of T-cell repertoire in allogeneic hematopoietic stem cell transplant recipients. *Bone Marrow Transplant* 2015;**50**:1227–34.
99. Choudhury NJ, Kiyotani K, Yap KL, Campanile A, Antic T, Yew PY, Steinberg G, Park JH, Nakamura Y, O'Donnell PH. Low T-cell receptor diversity, high somatic mutation burden, and high neoantigen load as predictors of clinical outcome in muscle-invasive bladder cancer. *Eur Urol Focus* 2016;**2**:445–52.
100. Inoue H, Park JH, Kiyotani K, Zewde M, Miyashita A, Jinnin M, Kiniwa Y, Okuyama R, Tanaka R, Fujisawa Y, Kato H, Morita A, Asai J, Katoh N, Yokota K, Akiyama M, Ihn H, Fukushima S, Nakamura Y. Intratumoral expression levels of PD-L1, GZMA, and HLA-A along with oligoclonal T cell expansion associate with response to nivolumab in metastatic melanoma. *Oncoimmunology* 2016;**5**:e1204507.
101. Park JH, Jang M, Tarhan YE, Katagiri T, Sasa M, Miyoshi Y, Kalari KR, Suman VJ, Weinshilboum R, Wang L, Boughey JC, Goetz MP, Nakamura Y. Clonal expansion of antitumor T cells in breast cancer correlates with response to neoadjuvant chemotherapy. *Int J Oncol* 2016;**49**:471–8.
102. Ikeda Y, Kiyotani K, Yew PY, Sato S, Imai Y, Yamaguchi R, Miyano S, Fujiwara K, Hasegawa K, Nakamura Y. Clinical significance of T cell clonality and expression levels of immune-related genes in endometrial cancer. *Oncol Rep* 2017;**37**:2603–10.
103. Kato T, Iwasaki T, Uemura M, Nagahara A, Higashihara H, Osuga K, Ikeda Y, Kiyotani K, Park JH, Nonomura N, Nakamura Y. Characterization of the cryoablation-induced immune response in kidney cancer patients. *Oncoimmunology* 2017;**6**:e1326441.
104. Kiyotani K, Park JH, Inoue H, Husain A, Olugbile S, Zewde M, Nakamura Y, Vigneswaran WT. Integrated analysis of somatic mutations and immune microenvironment in malignant pleural mesothelioma. *Oncoimmunology* 2017;**6**:e1278330.
105. Mai T, Takano A, Suzuki H, Hirose T, Mori T, Teramoto K, Kiyotani K, Nakamura Y, Daigo Y. Quantitative analysis and clonal characterization of T-cell receptor beta repertoires in patients with advanced non-small cell lung cancer treated with cancer vaccine. *Oncol Lett* 2017;**14**:283–92.
106. Saloura V, Fatima A, Zewde M, Kiyotani K, Brisson R, Park JH, Ikeda Y, Vougiouklakis T, Bao R, Khattri A, Seiwert T, Cipriani N, Lingen M, Vokes E,

Nakamura Y. Characterization of the T-cell receptor repertoire and immune microenvironment in patients with locoregionally advanced squamous cell carcinoma of the head and neck. *Clin Cancer Res* 2017;**23**:4897–907.

107. Kato T, Matsuda T, Ikeda Y, Park JH, Leisegang M, Yoshimura S, Hikichi T, Harada M, Zewde M, Sato S, Hasegawa K, Kiyotani K, Nakamura Y. Effective screening of T cells recognizing neoantigens and construction of T-cell receptor-engineered T cells. *Oncotarget* 2018;**9**:11009–19.
108. Vandiedonck C, Knight JC. The human Major Histocompatibility Complex as a paradigm in genomics research. *Brief Funct Genomic Proteomic* 2009;**8**:379–94.
109. Jin Y, Wang J, Bachtiar M, Chong SS, Lee CGL. Architecture of polymorphisms in the human genome reveals functionally important and positively selected variants in immune response and drug transporter genes. *Hum Genom* 2018;**12**:43.
110. All of Us Research Program I, Denny JC, Rutter JL, Goldstein DB, Philippakis A, Smoller JW, Jenkins G, Dishman E. The "all of us" research program. *N Engl J Med* 2019;**381**:668–76.

CHAPTER

Technology of clinical genomic testing

Judy S. Crabtree, PhD [1,2]

Associate Professor, Department of Genetics, Louisiana State University Health Science Center, New Orleans, LA, United States[1]; *Scientific and Education Director, Precision Medicine Program, Director, School of Medicine Genomics Core, Louisiana State University Health Science Center, New Orleans, LA, United States*[2]

Chapter outline

Introduction

Since the completion of the human reference genome in 2003, the focus within the broader genomics community has shifted to the identification of genetic variants that exist between individuals and populations. With the advent of next-generation sequencing (NGS), there is now a tractable method in terms of speed, accuracy, and cost, to sequence the DNA of many individuals to identify such variants. The availability of this technology has opened new doors for clinical translation of variant detection and the correlation of these changes with disease, drug responses and therapeutic outcomes. New goals by the community envision whole-genome or whole-exome sequencing being available to all patients and the results of this testing seamlessly integrated into the clinical decision-making process.

The Human Genome Project

The Human Genome Project (HGP) was an effort to sequence the human genome and to identify all the genes encoded by the DNA. The HGP was an international

Clinical Precision Medicine. https://doi.org/10.1016/B978-0-12-819834-6.00008-2

research effort that involved scientists from all over the world including the United States, the United Kingdom, France, Germany, Japan, and China. The HGP began in 1990 and was coordinated by the National Institutes of Health and the US Department of Energy. Completion of the human genome in 2003 resulted in what we now refer to as the reference genome, which is a compilation of DNA sequences from many anonymous volunteers from diverse populations. To ensure anonymity, volunteers replied to local advertisements around the areas where DNA laboratories were located, and 5–10 times as many individuals were recruited than necessary such that the volunteers would not know whether their sample was used. All identifying labels were removed from the blood samples before the sequencing was performed. The resulting sequence was the basis for identifying all the genes in the human genome and for supporting preliminary understanding of regulatory regions, noncoding elements, and repeated regions.

The HGP was sequenced using a technology called dideoxynucleotide chain termination, or Sanger sequencing, named after Nobel Laureate Frederick Sanger. This process uses DNA polymerase to incorporate nucleotides, called dideoxynucleotides (Fig. 8.1) that lack the 3′ OH required for DNA chain extension. The random incorporation of these nucleotides in a sequencing reaction terminates the

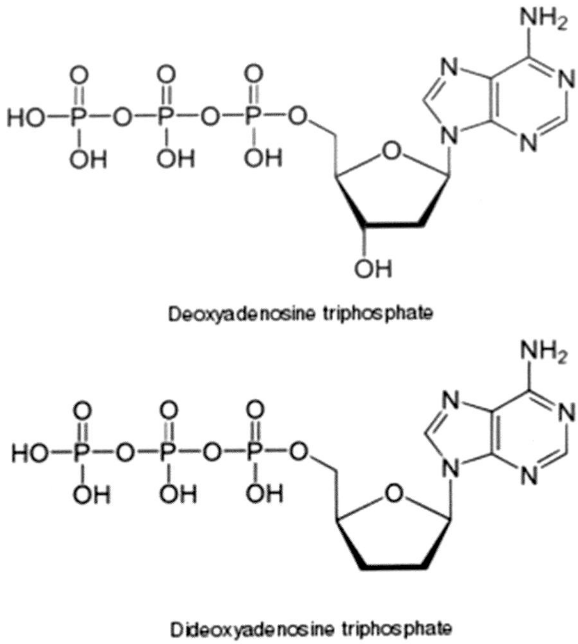

FIGURE 8.1

Deoxynucleotide versus dideoxynucleotide.

By Estevezj- Own work, CC BY-SA 3.0, https://commons.wikimedia.org/w/index.php?curid=23264166.

chains of growing DNA, generating a nested fragment set. The nested fragment set is separated by size using electrophoresis, and the sequence determined based on the mobility of the individual fragments. Over time, new technology and instruments were developed to facilitate detection of these fragments—early DNA sequencing was performed using radioactive labeling of nested fragment sets, and later technologies utilized four different fluorescent groups with one added to each of the dideoxynucleotide terminators for A, C, G, and T (Fig. 8.2).

To perform sequencing of the entire genome, the chromosomes had to first be broken up into smaller, manageable pieces using a process called hierarchical shotgun sequencing (Fig. 8.3, left panel). Genomic DNA from the human donors was first broken up into pieces of 150–350 kb and cloned into vectors called bacterial artificial chromosomes (BACs). The early mapping phase of the HGP used genetic (relationship of genes to each other based on how likely two gene regions are to be inherited together by the next generation) and physical (physical distance between DNA sequences on a chromosome) mapping to order and identify the location of the BAC clones on the original chromosomes. Genetic mapping determines the relationship of genes (or gene regions) to each other based on the probability that the two genes (or gene regions) will be inherited together in subsequent generations. Physical mapping is the actual physical distance between genes or gene regions. Physical maps were determined using restriction enzyme digestions of BACs,

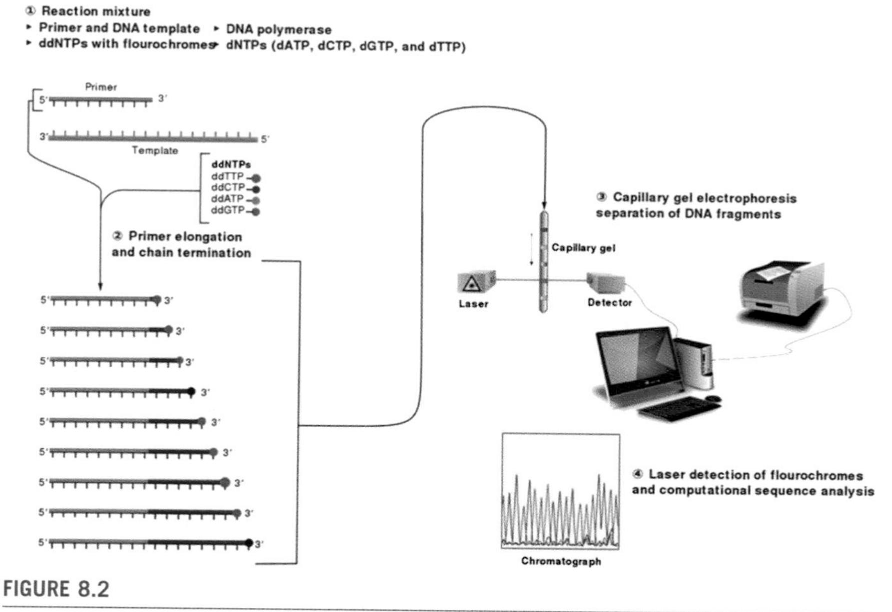

FIGURE 8.2

Sanger sequencing.

By Estevezj - Own work, CC BY-SA 3.0, https://commons.wikimedia.org/w/index.php?curid = 23264166.

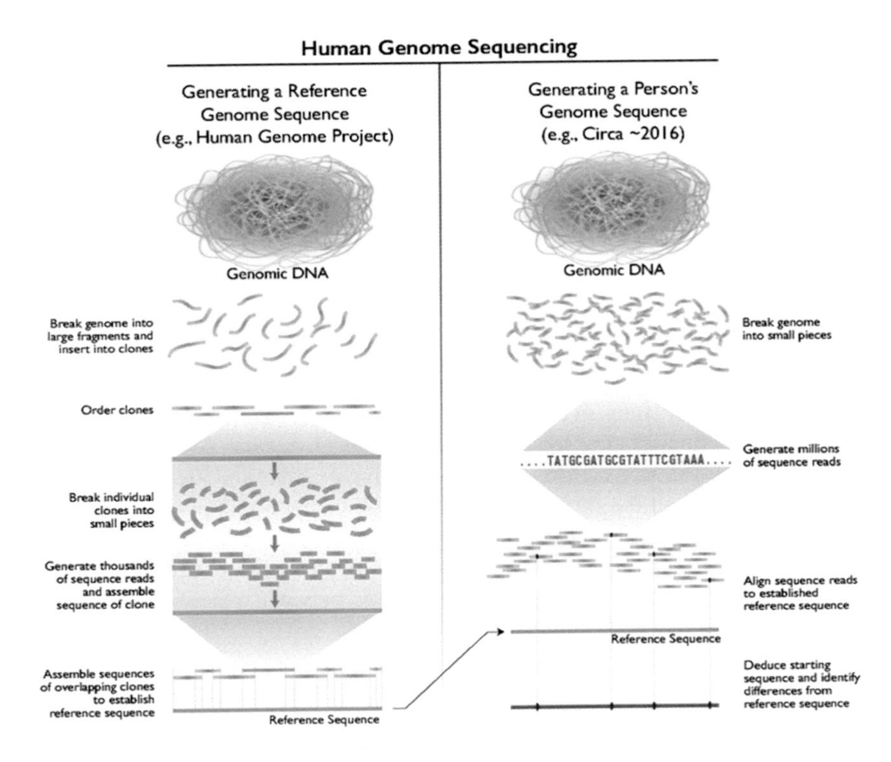

FIGURE 8.3

Human genome sequencing.

https: //www.genome.gov/about-genomics/fact-sheets/Sequencing-Human-Genome-cost

separating the fragments based on size to get a "fingerprint" of the BAC, and finally piecing together the overlapping fingerprints to determine the order of the BACs.

By selecting BAC clones that were predicted to have overlapping ends, a minimal tiled set of BACs that would cover each chromosome was then selected for sequencing. Each individual BAC was further fragmented into 500–1000 bp pieces and cloned into plasmid vectors, each of which was a manageable size and sequenced using Sanger sequencing. The individual plasmid sequence reads were assembled based on overlapping sequence information to generate the entire sequence of each BAC, and then the sequences of the individual BACs were concatenated in the same way to compile the complete sequence of each chromosome. The complete collection of all of these sequences is the reference human genome.

The HGP took 13 years and $2.7 billion in fiscal year 1991 dollars. Toward the end of the HGP, competition from the private sector encouraged technology development and increased innovation. By the time the reference genome was complete, new technologies were emerging that would revolutionize the DNA sequencing world as it moved toward the goal of clinical translation of DNA sequencing and improved clinical care. Researchers and clinicians were beginning to ask new questions with foundations in genomics, and the technologies had to become more approachable in terms of cost, time, and accessibility if they were going to be clinically useful.

Next-generation sequencing

The revolution in genome technology resulting from the HGP led to a number of new approaches using massively parallel, or deep sequencing, the so-called NGS technologies. Several platforms emerged in the early 2000s with different approaches to template preparation and chemistries, but the platforms all shared the common concepts of miniaturization, amplification of discrete DNA molecules, and detection of single nucleotide additions within a flow cell of the size of a standard microscope slide. NGS allows sequencing of millions of small fragments of DNA simultaneously, resulting in gigabases (Gb) of sequencing output per instrument run. These small sequencing reads are pieced together using high powered computer programs, or bioinformatics, and are aligned based on the reference genome into the complete sequence of the sample. Using these platforms, each nucleotide of the human genome is sequenced many times, providing unparalleled sequencing depth and accuracy in variant detection.

NGS can identify many more variants than traditional Sanger sequencing. As discussed in previous chapters, variants can fall on a spectrum from small single base changes to large insertions/deletions to large chromosomal translocations/inversions. Historically, Sanger sequencing has been used clinically to test for single base changes or to identify small insertions/deletions, whereas fluorescent in situ hybridization (FISH) or comparative genomic hybridization (CGH; see Chapter 1) has typically been used for larger chromosomal rearrangements and copy number changes. NGS can provide information on all of these rearrangements with a single experiment, providing the full spectrum of genomic variation in a given patient sample that can be derived from blood, saliva, or a tissue biopsy. NGS provides unbiased results for genomic variation and can identify unknown, completely novel variants that may be the basis for genetic syndromes of unknown etiology, changes in tumor behavior to treatment, or mosaicism, all of which would be missed by traditional, region-of-interest Sanger sequencing. Furthermore, the depth of coverage possible with NGS allows for identification of low-frequency variants and sequencing from samples with low abundance such as detection of fetal DNA from a maternal blood sample or DNA from circulating tumor cells in the blood of a cancer patient. The primary disadvantage to NGS is the challenge associated with identifying variants in hard to sequence regions, i.e., those with high guanine/cytosine (GC) content or repeated regions, such as those in repeat expansion disorders such as Huntington's disease, fragile X syndrome, or spinocerebellar ataxia, type 1.

Whole-exome sequencing

Genes within the genome consist of several identifiable parts including the promoter region, exons (or the portion of the gene that codes for a protein), introns (noncoding regions of the gene in between exons), and untranslated regions (also called 3′ or 5′ UTR depending on their location with respect to the start of the gene) (see Chapter 2). The exome consists of all the exons in the genome, which is approximately 180,000 exons, or about 1.5% of the total 3 billion nucleotides of the human genome. Interestingly, variants in the exome are responsible for more than 85% of single-gene genetic disorders. Whole-exome sequencing is significantly less expensive than whole-genome sequencing, generates 1/15 the amount of data (big data that have to be stored and linked to patient electronic medical records), requires less computing power for analyses, and can detect many pathogenic variants. The process of whole-exome sequencing uses the same NGS sequencing technology as for whole-genome sequencing but includes some additions to the upstream sample processing to isolate and capture exonic sequences (Fig. 8.4). Typical workflows include selection of exon-containing sequences using biotin-labeled RNAs that are complementary to exonic sequences and purified using streptavidin beads. Purified, enriched exonic fragments are then subjected to NGS just as in whole-genome sequencing. The technical challenge with whole-exome sequencing is that the sequencing results are only as good as the RNA capture library. Development of these libraries by a number of companies has improved the quality of exon capture in recent years, covering regions of the genome that are difficult to accurately capture due to GC content, and including regulatory regions and UTRs that may contain

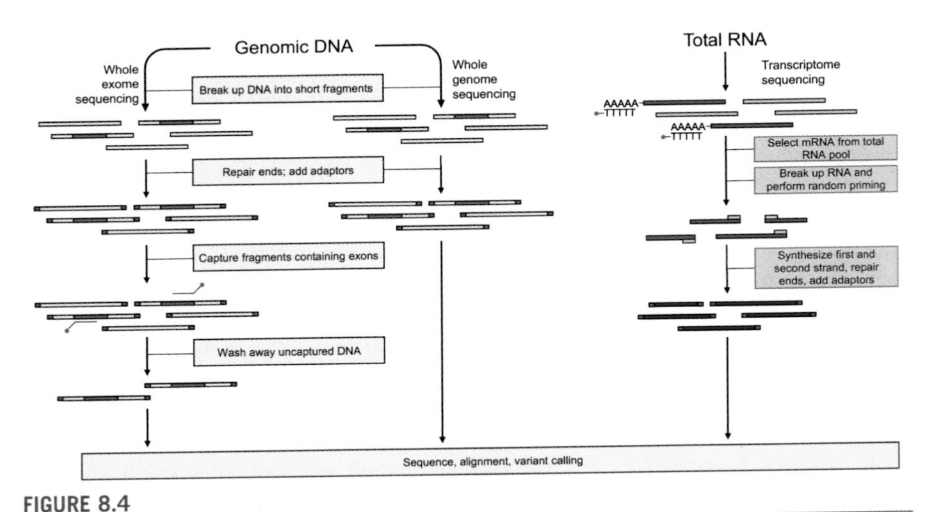

FIGURE 8.4

Workflow comparisons between whole-genome, whole-exome, and transcriptome sequencing in the era of next-generation sequencing.

Bras, Guerreiro and Hardy 2012 Nature Rev Neuroscience 13: 453–464, Figure 2.

variants. Most libraries contain capture RNAs for all known genetic variants that impact the expression of proteins within the cell.

Transcriptome sequencing is also an option, although this approach is more often used in the research setting (Fig. 8.4). DNA is transcribed into RNA, and during splicing and processing, a polyadenosine tail is added to the mature mRNA. The poly A tail can be used to capture only the mRNAs in a sample and then used to generate complementary DNA (cDNA). Once double-stranded cDNA is generated, it serves as the template to be sequenced by NGS technology. The goal of this approach is to interrogate only the regions of the genome that are transcribed into RNA.

Panels

An alternative to whole-exome sequencing is a focused panel, which contains only a discrete number of genes to be interrogated, and yields high coverage of relevant genes of interest. There are a variety of chemistries and protocols to enrich for the genes and regions of interest (as with the biotin probes for exon capture above), and the approaches vary depending on the platform in use. Focused panels exist in support of many disease states including epilepsy, cardiovascular, neurological, endocrine, and oncology—with subpanels for hematologic malignancies, breast/ovarian, colon/Lynch Syndrome, lung, and gastrointestinal cancers. As the price of whole-exome sequencing decreases, one can envision the scenario where whole-exome sequencing overtakes the smaller panels yielding more information for similar or lesser cost.

Single-gene resequencing

A final option in the clinical setting is single-gene resequencing when variants are suspected in a known gene linked to disease. This is most often performed by the lower throughput Sanger sequencing for known Mendelian disease genes such as BRCA1/2 (breast/ovarian cancer), MEN1 (multiple endocrine neoplasia, type 1), DMD (Duchenne muscular dystrophy), or CFTR (cystic fibrosis).

ELSI and GINA

From its inception, the HGP also included a program called Ethical, Legal, and Social Implications (ELSI), to anticipate, understand, and address the implications of genetic and genomic research on individuals, families, and communities.[1] The ELSI program, still in existence at the National Institutes of Health, now solicits and funds investigator-based research on the impact of genomics research on genomic medicine, legal and policy issues, and broader societal issues (https://www.genome.gov/Funded-Programs-Projects/ELSI-Research-Program-ethical-legal-social-implications).

The Genetic Information Nondiscrimination Act of 2008 (GINA) is a federal law in the United States to prohibit discrimination on the basis of genetic information with respect to health insurance and employment (https://www.genome.gov/about-genomics/policy-issues/Genetic-Discrimination). Health insurers may not set premiums, determine coverage, or eligibility based on genetic information, and are prohibited from requesting genetic testing (or requiring individuals to provide results of such testing) prior to health insurance-based decision-making. GINA also includes provisions to prohibit employers from using genetic information to make hiring, firing, promotion, pay grade, or job assignment decisions. However, there are some cases where GINA does not apply, for example, it does not apply to companies with fewer than 15 employees, and the law does not include any protections with respect to long-term care insurance, life insurance, or disability insurance.

Conclusions

Testing in the clinical genomics laboratory has changed significantly during the years since the HGP and the technology continue to advance toward the ultimate goal of understanding the role of genomic variation in the cause and treatment of disease. As genomic testing matures and moves into the clinic, additional challenges such as insurance, Medicare and Medicaid reimbursement, access to care, data security, and increased privacy of personal health information will need to be addressed.

Reference

[1] McEwen JE, et al. The ethical, legal, and social implications program of the national human genome research institute: reflections on an ongoing experiment. *Annu Rev Genom Hum Genet* 2014;**15**:481–505.

Index

'*Note*: Page numbers followed by "f" indicate figures, "t" indicates tables.'